NO CONDITION
IS PERMANENT

NO CONDITION IS PERMANENT

RISK, ADVENTURE AND RETURN
The Business of Life

SIR BOB REID

FONTHILL

www.fonthillmedia.com
office@fonthillmedia.com

First published in the United Kingdom
and the United States of America 2020

British Library Cataloguing in Publication Data:
A catalogue record for this book is available from the
British Library

ISBN 978-1-78155-803-4

Typeset in 10.5pt on13pt Sabon
Printed and bound in England

For Jo and the boys

CONTENTS

1

Early Years

My father's shop in the high street had full-length windows on each side of the front door where huge sides of beef and whole sheep hung on steel hooks. Behind them stood the wooden benches where the primary butchering was carried out. The shop's four butchers each carried a knife sheathed in a stout leather pouch, and a steel to hone the blade. Marble counters faced the customer, and the floor was lightly sprinkled every day with fresh sawdust to absorb any slippery grease. There were two rooms below the shop: one with pens for poultry and the other with a sink for washing pie dishes and trays, a large slate table and an industrial mincer. It was a hive of activity. At nine years old I was as proud that my father owned the shop as I was determined to be involved.

I already caught, plucked and cleaned out hens, took and delivered orders and collected the money. Plucking was easy, provided the hen was warm, because then the feathers came out without difficulty. The only irritation came from the fleas that hopped out of the plumage and onto your clothes, and from there onto the back of your neck. It was a strange feeling, slapping yourself to crush the minute creatures against your skin, but any qualms you had about applying the fatal blow disappeared with the itch.

We usually sat in a circle to pluck, and when we were busy we were joined by an old contract plucker who moved twice as fast as any of us. He rarely spoke, but was watchful and would occasionally give a hand with the difficult wing feathers. As he worked, he collected the hens' coloured identification rings on his table. To my mother's distaste I had taken to wearing them on my fingers. I felt they signalled to the world that I was a working man who could pluck hens for a living.

That summer I had been up every day at half-past six. I loved being part of the team—doing something useful and even getting paid for it! Late one afternoon, I was helping prepare suet for pies when the mincer I was in charge of gurgled to a halt. Alone in the room, I thought it would be a simple matter to clear the blockage, so I reached forward into the machine. It restarted with a jolt, gripped and bit into my fingers, and pulled me down, the blades chewing upward towards me. My mind was engulfed with pain. I could hear the mincer stutter and protest as chips of bright hen rings pinged out into the basin. As the grinding slowed and finally cut out, I remember thinking I was going to be in big trouble. One of the butchers rushed down to see what the commotion was about and yelled for my father. Within minutes I was in his arms and on my way to Dundee Royal Infirmary. I can remember his face set against the street lights as we sped through the city. That evening a New Zealand surgeon trimmed my arm and sewed it up, rather like a pot roast.

The next morning I woke in a bed I did not recognize. I opened and closed my eyes many times, praying that this was a bad dream. Soon, I told myself, it would be time to go to school, so I had better wake up properly and stop this horror nonsense. But reality persisted. A nurse arrived to wash my face and sit me up.

Her next job was to fetch me some breakfast because, she said, I must keep my strength up. I was beginning to realise it was not my strength I had lost but my hand, and no matter how much breakfast I had it was not going to grow back. My ward-mates were not much help either. There was an old man fast asleep on the left and a young man with a huge bandage on his head on the right. I was told a slate had fallen off a roof and hit him; he had hardly recovered consciousness and had not spoken. The staff were obviously very worried about him and this made me feel a bit less miserable—though guilty that somebody else's accident should make me feel better about my own. *After all*, I thought, *I'm not dead—unless this is Hell. It certainly isn't Heaven.* I reasoned that the nurse, who looked a happy sort of person, certainly would not be in Hell … but I made up my mind to check with her all the same.

'Where am I?' I said.

'This is the Dundee Royal Infirmary.'

'Does the King come here?'

'No.'

'But why is it royal if the King doesn't come?'

She replied, 'You need your breakfast.' This seemed an unsatisfactory answer, but I left it there.

Breakfast came. The Cornflakes were good and crunchy, but then came the boiled egg. I felt it; it was warm. After a few awkward left-handed taps with the knife it was clear that the top was not going to come off. I looked to one side; the old man was still fast asleep. On the other, Slates was unconscious, so I shuffled to the edge of the bed and felt for the frame. To my relief, it was iron. Taking the egg in my left hand I crunched it against the metal. It was not soft-boiled, thankfully; no warm liquid ran through my fingers onto the floor. But I had done it. I ate the egg and some shell with great satisfaction.

By this time everyone in the huge ward was beginning to wake. Nurses appeared, rushing about tidying beds, taking temperatures and marking up charts. I asked them why they were doing this in my case since I did not have a cold and had not even sneezed. I could not make the connection between losing a hand and having a problem with my temperature. At this point a woman in a dark-blue uniform arrived with a watch pinned to her breast. Her starched hat was so white it seemed to sparkle. She clearly struck fear into the nurses, especially the nice one who could not answer my questions. *She must be bottom of the class*, I thought sadly.

It was to Nice Nurse that Sister directed her queries. 'How is he?'

'Very nervous and asking a lot of strange questions.' Sister advanced to the bedside and stared at me severely before saying with great assurance, 'You're going to be alright, you'll see.' This I thought was a really stupid comment, no matter how well intended. I was not going to be the same again—I was a hand short and there was a piece of eggshell on the floor to prove it (she was aware of that, too: she had commanded Nice Nurse to pick it up). After a smile, which I returned, she moved on to Slates, and I was interested to hear how badly he was doing. They were about to take off the bandages and apply another dressing when the doctor arrived. I braced myself. The dread of seeing a crushed skull was exciting. Would I be sick, I wondered, if I saw his brains? Were his brains like the brains in the butcher's shop? I would need to take notes to tell my brother, who wanted to be a doctor—*but wait*! I remembered I no longer had the damned hand I used to write! Still, there were plenty of left-handed writers, I recalled—Sandy Merilees for one. And if he could write left-handed, I certainly could.

The doctor came to my bed. They had done nothing with the old man on my left. I wondered if he was dead. Sleeping in a place with a dead person on one side and a man with a cracked skull on the other was not too good. I asked the doctor if the old man was OK. He asked if I knew him. This was another daft question. I thought, *How could I know him? I arrived in the middle of the night, I never spoke to him and he never stirred.* The unreality of everything—the cleanliness, the starched and purposeful nurses, the doctors in white butchers' coats—seemed totally bizarre.

After a whispered consultation the doctor, who had a friendly face, asked if I'd like to see my arm. Compared with looking at my neighbour's cranium, that would be easy. 'Yes, of course,' I said rather briskly.

'It won't surprise you too much?' he asked gently.

'No,' I said. 'My hand has gone.' So they took off the bandages and there it was, my stump, my friend for life, the focus of interest of all my children and grandchildren. It looked strange, and did not seem part of me. 'Will it grow?' I asked.

'Above the elbow, yes, below the elbow, no. Fingers don't grow again, you know.'

This man must really think I'm thick, I thought. *Doesn't he know I'm a plucker, and I can skin a rabbit, and recognize the heart, the liver, the kidneys and the gizzard? I think I'll be a doctor—I know one body part from another and with training I could be better than they are.* As the doctor then laboured over my temperature chart—which recorded one solitary reading—I was sure of it. *But does he fancy my bottom-of-the-class nurse?* I wondered. *Oh, boy! There's a lot going on here—wait till I see my brother to tell him!*

As they moved on to Slates my throat tightened and my heart began to race. But then they pulled the screens around him and stayed in there a long time. When the doctor re-emerged he gave me a wink. It was beginning to look like Slates was a good excuse for huddling behind a curtain with Nice Nurse.

The following day my parents turned up. This was sad. My mother had obviously been crying and could still hardly keep her lip from quivering. My father too, uncharacteristically, was almost upset.

My father soon regained control and started to talk about writing, how long it would be before I could go home, and then people who had lost hands in the War and how they had done. He gave me a long list of people who had asked after me and sent their best wishes. Some of those included

surprised me—some I could not believe, and others I did not want to
believe. Teachers asked after you only because they wanted you back to
torture you. Ministers only wanted the Sunday School numbers made up.
Sandy Watt was different: he would care. Sandy Merilees with his left-
hand writing was now a vital ally and he was there, too. I was relieved
when my parents left. It had been hard watching them trying to restore
what they could not.

Slates disappeared in the middle of the night and the old man woke up
at last. The nurse appeared and told me bewilderingly that I had to learn
to walk again. This did not amaze me—anything was possible in this stark
theatre of nightmares. Even today they were pushing beds out onto the
veranda. The sun was out and patients were taking the Dundee air.

My legs did not belong to me. They felt weightless and I did not seem to
be able to place them one after the other. *Oh God, this must be connected
to my arm*, I thought. *You lose a hand and you can't walk!* My father had
not said anything about this.

'It won't take long,' the nurse said reassuringly, and put her arm around
me. She was warm and smelled nice and clean. Carbolic soap and perfume
seemed a good mix. Up and down the ward we went until my legs were
back under control. I was off—side effects: nil. This was great.

But there was the problem of the stump. This was becoming an
addition, not a subtraction. Very few people—and no one in my school—
had a stump. How long would it take to get well? I could feel my fingers
quite clearly; was this bad or good? I held it up bent—was this right or
wrong? It did not seem part of my body—would it become so? I saved up
these questions for the doctor and he did well. He liked sitting on my bed,
and he looked at me when he spoke.

The lavatory was at the end of the ward and on day three I was allowed
to go on my own. The cubicle had a heavy panelled door and metal locks
to forestall intruders. The bodily function was not a problem but cleaning
up was more difficult with an inexperienced left hand. This meant more
paper and more practice. I was not worried about that in the privacy of
this small room with its mottled-glass window, but then a problem did
arise: getting out. Try as I might, I could not open the door. I took off my
slipper and whacked it, but to no avail.

One thing was certain: I was not going to shout for help. If I did I would
never have any privacy as long as I was in that place. It dawned on me
there was another way out: the window. I quietly unfastened the polished
catch and saw that it opened close to the veranda. It would not be difficult,

I thought, to scramble onto the ledge, lower myself onto the balustrade and from there hop to safety. The time had come to make my move so up I went, over the top onto the railing and then onto the veranda. To the surprise the patients taking the sun in their beds, I skipped past them into the ward.

It was a matter of minutes before Sister and Nurse confronted me with the enormity of my action. We were three floors up! I could have fallen to my death! My arm was only mending and I could have damaged it terribly. I might have to stay in the infirmary for weeks. I got it all, and no smiles. When they threw in the final question—'And how are we to open the cubicle now?'—my chance as I saw it came.

'I can get back in,' I volunteered.

'How?'

'The same way I came out!' This, I realised immediately, was a bad answer, so I tried to look contrite. I think they thought I was going to cry. Sister disappeared and Nurse gave me a doughnut, saying: 'You are going to your own hospital tomorrow. They have a nice bed for you and your aunt will be there to look after you.'

I liked my aunt. She was senior sister at the hospital and I knew from experience she would take no nonsense. It was good news. 'Mind and behave,' she said. And with a twinkle and a smile added, 'Don't worry: your ward is on the ground floor.' I couldn't resist: 'That's good, Sister, but actually the whole hospital is on the ground floor.'

*   *   *   *   *

It was with mistaken relief that I returned to the cottage hospital in my home town, Cupar, to spend ten days convalescing. I realised then how life had changed because of this stupid, stupid mistake. Aunt Chris was highly efficient, her aprons gleaming white and so starched she could step into them. Her watch was always on time and she had medals. She had the character and will to make the first move. Tears and sympathy were over. It was down to business.

I could no longer rely on my hands for a living because I had only one. My butchering career was over. All the skills I had acquired—drawing chickens' necks, delivering messages, mincing—were irrelevant. From now on I would have to work with my brains. The first imperative was to learn to write with my left hand; this was to be achieved before I left the hospital. Morning, afternoon and evening saw me hard at work, and when the time came I had done it. In ten days I had built the skill to write: not beautifully, but adequately. As it was to be brains, it was back to school—quickly. This was the difficult bit. I had become unusual, a sort of freak who could be talked to but not played with in case I started to bleed. This gradually wore off but it put a shell on me which I would never lose. Being odd either forces you back into your shell or makes you learn to live with one; you take it with you and use it to forgive yourself for your ineptitudes and your stumblings over simple things.

The chief medical officer for Fife, whose meat I continued to deliver, had me in his sights. He decided I should be fitted with a prosthesis. I asked my brother Jim what that was and as always he had the answer: an artificial arm. So the process began. I was to go to Musselburgh, a military hospital, for ten days, and be measured and fitted. I was not to be visited for a whole week. Now aged ten (double figures!) I found this an amazing experience. It was December, so Christmas was coming and the place was already decorated. The hospital was full and everybody had lost something—legs, arms, eyesight. The soldiers were not allowed regular visitors either, it seemed.

The day was a dream. We rose at seven, trooped in to breakfast at eight and then went to play in the gym. We did exercises, caught balls, hit clay sticks at wall targets and played football and handball. In the afternoons we played cards. I knew how to play whist, so was included. Every evening there was entertainment, but I was given the job of pushing a legless soldier three miles down the road to Musselburgh, where he was attending tapestry classes, and back again. This was an education. He sent me to the cinema while I waited for him—upstairs, as assistants were not allowed downstairs. On the way back to the hospital he told me how he had lost his legs in a commando attack; it was so exciting that we were back in no time. The next time he had lost those same legs in an aeroplane crash, and the next after having been trapped in a submarine. This was better than the movies.

At the cinema I eventually moved downstairs where the seats were cheaper and you were nearer the action. One evening just after the cowboy film had started I noticed a disturbance in the aisle. Someone was being levered out of a wheelchair into a cinema seat. It was my man! He had played truant on his classes—if they ever existed—and was obviously with a close friend, since he spent the evening embracing her with some enthusiasm. I slipped upstairs before the end of the film, then there he was, waiting to tell me about his escape from a burning Spitfire. I liked him. He was a good friend and despite his leglessness was always happy, but I fear his walls will never be hung with tapestry.

When a week had passed my mother came to visit. She was quick to see I had not bathed for seven whole days and that there were black rings around my eyes from late nights. But what she did not see was that this hospital was not about artificial limbs; it was about realising you were not alone. Being younger than the rest, I was more adaptable and so less concerned about the future. After all, this was where brains would come in—and they were still there as far as I knew.

*   *   *   *   *

The dreaded apparatus eventually arrived. It had a steel sheath with straps that went across my back and around my good shoulder. Onto the sheath you could fix a solid hand, which you could twist around like a machine gun, and it had a thumb you could lift and release at speed. To my great joy there was also a hook and a strange attachment for cleaning your teeth.

The hand was fantastic. Turning it to face the bowler at rounders, I sent successive balls towering out of the playground. The hook did not make it to Day Five as a note from school advised my mother that I had torn three shirts at the collar and frightened all the girls. Three weeks later after a brawl and a knockout, the hand went the same way— that left just the toothbrush, and who wants to wander around with a toothbrush sticking lonely out of your right sleeve? I was one-handed and left-handed, and cosmetically the apparatus was not going to enhance me.

The episode of the artificial hand was over, but the war was not. We were convinced we would be invaded by the enemy and had to prepare ourselves. My mother worked in the military command office and my father, ineligible for military service because butchery was a reserved occupation, was part of the Home Guard. He spent all his spare time training for resistance, and that is how I learned about Sten guns, Lewis guns and rifles. Jim and I were particularly fascinated by one of his training booklets, which described how you could thwart invaders at home. Following its instructions, we secretly chipped away at the mortar in the attic chimney breast until we could remove a brick and squeeze a hand through. The plan would then be to drop a grenade into the room below. I knew what had to be done, and my hand was small enough to do it. We thought about trying out our defence strategy on Uncle Jimmy next door, but decided that might not be an approved use of scarce ammunition—and perhaps an actual explosion in the home was not the best idea.

We had a Polish officer staying with us. He was an unhappy, resolute man missing his family and his country, apparently unlike the mad Frenchmen who were running around the countryside climbing trees, fording rivers and digging trenches.

The Poles had a bugler who played from the tower of the corn exchange every Thursday. We felt he was a very poor musician, or perhaps just a trainee. Eventually we persuaded our teacher to talk to his officer and ask whether he could be replaced with someone who could actually finish a tune. The reply amazed us. The bugler, like his colleagues, was from Krakow, which in the Middle Ages had been attacked by a horde of Barbarians. The town bugler sounded the alarm, but before he could finish he was shot in the neck and killed. That is why the tune never ended; it just faded out as it would have when the trumpeter fell. This sad story satisfied the class, although there were many ancillary questions for the officer, mostly technical but some about his name, age, even the colour of his hair ... but not his address. We left that to the teacher.

Our cosmopolitan society also included Italians, but they were locked up as prisoners of war. At least most of them were, until the Fife farmers, remembering what they had learned at school about the building of Rome, went to look among them for masons. There were quite a few, and to keep them occupied they were let out on day-release. The steadings (farm buildings) of central Fife still bear testimony to the excellence of their craft.

We were all busy learning our plane-recognition charts, building fortifications, looking for spies and collecting all sorts of things for the war effort. Conversations were all about the war. Our friends who ran the fruit farm were, like my father, First World War veterans. As the talk took a serious turn, Joe would declare that if the Nazis came down the road (gesturing out of the window to the road, which as far as I could see only came up—there was no way he could see the down, but when he was in full flow there was little point in interrupting him on detail, particularly as the good bit was coming ...) he would—with regret—shoot his wife and two girls and then fight to the end with his son, Angus. Angus liked the bit about doing away with his sisters; they had always plagued him. The rest he was not too happy about.

It was at Joe's fruit farm we spotted a Messerschmitt and reported it. We were officially congratulated, and celebrated with a parade and a display of weapons. The prize exhibit was the draper's son's airgun. He told us would save his live ammunition for the enemy but would demonstrate the gun's power with one carefully selected rowanberry, which he proceeded to load. To our wonderment he shot at short range into the plump bare thigh of one of our brigade. We gathered round and gazed at this red berry embedded in his flesh. It was beautifully flush with the rest of him. As the yelling died down, the enormity of the escapade grew.

'You'll need to get it out,' Jim remarked knowledgeably. 'It's a seed and it'll grow into a tree. You can't go walking around with a tree in your leg.' The observation produced more tears.

Although I was by far the youngest, they turned to me. 'You've been in hospital; do you know the surgeon?' I did, but I doubted this wound would warrant Mr Scrymgeour's talents. I mentioned his name to help. 'We want a doctor, not a doorman!' came the indignant reply. There was little point in explaining the niceties of medical nomenclature to these uninformed individuals. Instead, I offered a nail file, and with three of us sitting on him we Scrymgeoured the berry out of the target's leg. He gave no thanks and departed tearfully up (not down) the road.

At the next get-together, we all dutifully admired the single beautiful apple that the farmer's wife's special apple tree had produced. Later, as I happily tucked into a beetroot, apple and potato salad, the daughter burst into the kitchen to announce 'The apple has gone!' There was silence. Accusation was rife and I, simply for taking a second helping of the salad, seemed to be the prime suspect. I protested my innocence and bit my lip. If I cried that would suggest guilt. The identity of the thief remained unresolved until 57 years later, when the son of the house owned up to the dastardly deed. He confessed to my brother that he had taken the apple, eaten and enjoyed it. That was a long time to live with guilt for such a heinous crime.

I suppose if your child has been handicapped you are inclined to support his or her every endeavour no matter how pointless. And so it was with me. Without natural speed and fluency of stride it is difficult to make a success of running, but the challenge of the first Highland Games after the War was irresistible.

The boys' under-15 two-mile marathon was the target. At 12 years old I was too young, but this did not deter me. I trained practically every day until making the distance was not going to be a problem. Spurred on by my brother, a dedicated bibliophile whose reading time far exceeding my training time, I read all I could about running. It was pretty boring stuff as it was all about winners and past victories, and I had yet to make it to the starting line. As the day approached, Jim lectured me on the dangers of marathon running. He produced pictures of the exhausted Olympian Dorando Pietri who, in the last lap of the 1908 marathon, was helped up by umpires four times before finishing in first place. Because he had been assisted he was disqualified. Jim painted a lurid picture of heat exhaustion and dehydration, ignoring the fact that my race was two miles, not 26, and was in Cupar, not London.

But when the day came he was right: it was the hottest day in Scotland for 20 years. By the time I had changed, the first-aid people were already dealing with fainting spectators. The race was four times around the track. There were 25 competitors, half of whom wore spikes, something I had not previously encountered. On discovering my age the starter began to query whether I could go the distance—but I skipped past him onto the track. There seemed to be hundreds of spectators. We lined up and set off.

Mindful of Dorando, I settled down to what Jim subsequently described as an unnaturally fast walking pace. As the race wore on runners dropped out, and half the field had already gone by lap two. It was alarming to see them lying in exhausted heaps being tended to by the first-aiders. By the third time around I had been lapped, but another eight boys had staggered off. As I ran past on my last lap I took a good look at the finishing pair who had collapsed across the line. A steward strode towards me shouting, 'Finish! You're there!' What effrontery. I had not come to win a conceded victory. I had trained for the race and I was going to finish properly. My Dorando strategy had left plenty in my legs and to the delight of the crowd I dodged another white-coated steward halfway round. As I sidestepped a third he bellowed, 'We're starting the bike race, you daft little bugger!' but I ran on and crossed the line amid loud cheers, swerving onto the side of the track as the bikes swept by for their second circuit.

As I made my way to the tent I was pompously advised that I had been disqualified for failing to obey instructions. Quick as a flash, Jim, a future QC, defended me like the hot property I was not. 'He was not competing,' he said to my amazement. 'Money is involved, and we do not wish to endanger his amateur status. This was a chance for the home crowd to see him in action,' he informed the dumbfounded official. We left in high spirits, my brother delighted with his riposte, and me aware that this was the high point of a very short athletic career.

However, the afternoon's excitement was not over. I then set off with my young girlfriend to the fair close by. First stop was the shooting stall where I could demonstrate my skill as a marksman.

Stallholder Mr Balsillie was a large, strong man with tattooed biceps. He stood sweating in the sun, breaking guns and loading them with four darts each. It was all in a good cause. He took my threepence and I took my gun. My first shot went well wide as the trigger was stiff, and pulling it took the rifle off-centre. With my second dart I took careful aim and it happened again—but this time it sped unerringly into the heart of Mr Balsillie's tattoo.

It stayed there rather like a decoration—a crown sergeant's pip. Balsillie's howls of rage and pain as he withdrew it attracted a growing crowd, and his 'quiet man' reputation began to look doubtful. When I drew his attention to the fact that I had two darts left, he seemed less than interested and snatched back the gun. The next day I went to his house to apologise, but was quite relieved to find he was not there. His wife reassured me that he had recovered, and gave me sixpence to cheer me up.

The Balsillie incident led to long discourses at home on the importance of the customer. The idea that 'the customer is always right' sat uneasily with Jim, who invariably thought they were wrong, and what's more they should be told as much. Customers should be graded by intellect, not by their capacity to buy, was the core of his argument. As the spirited debate raged, I delivered the meat to a bewildering array of people, some of whom were polite, some of whom were plain rude and some just not interested. But among this bunch there were jewels.

One lady was convinced that her next-door neighbour was going to do away with her precious cat. Every time I found the cat she gave me a tip, but it was not a difficult search as the cat had an inexplicable attraction to the neighbour's house, much to the lady's annoyance. Calming her down was time-consuming, but financially rewarding.

Then there was the colonel on the hill with the Lagonda and the beautiful companion. It seemed a world away as I walked up the hill and he swept by in his open-topped car next to this film star, her scarf floating in the wind. I fantasised about this woman but nobody knew her name. Sixty years later quite by chance and to the surprise of both of us I met her daughter, and she was just as beautiful.

But the real jewel I left to the end of my round. Mrs Henderson was 92. She was lonely, she always had coffee and a chocolate biscuit ready for me, and I never left before half an hour had gone. Gradually she told me more about her life. As a little girl she had seen one of the last public hangings in Scotland. Three men had been accused of murdering the Post Office manager when he disturbed them in the middle of a robbery. The third man, the lookout, was really only a boy, and on the gallows that day he looked frightened and childlike. In front of the huge crowd that had assembled to watch, both the boy's co-accused pleaded for the youngster to be spared. Mrs Henderson, then aged only eight or nine, was with her parents on a beautiful summer's day, and she sobbed her heart out as the rope went around the boy's neck, and he fell and dangled there. A brutal headline from the time had read 'It's a braw day for hanging!' This traumatic event had made Mrs Henderson an implacable abolitionist.

In telling the story she was weaving her magic. She made a 12-year-old think about things he would never otherwise think about. She made you take a view. Her eldest son had won all the school prizes, taken a First at university and risen quickly through the ranks of the Civil Service. Tragically, he had died on a mountain climb in the Cairngorms. She said that part of her life had stopped that day and she still had not come to terms with it, but her late son's interest in world affairs had left her with an insatiable curiosity about a huge range of subjects.

As the weeks went by Mrs Henderson's reminiscences turned to advice about life and the future. She was concerned about friendship, partnership and marriage. A young boy absorbing romantic instruction from a 92-year-old is, looking back, bizarre, but I was hooked on this woman. She talked to me as no one else did. She warned me about good looks and superficial attraction; she wanted me to find a deeper relationship that would last. Marriage must be of the mind, she said, because that is enduring. This put my film-star Lagonda woman in her place and I even felt guilty about the thoughts I had had about her. As two years passed these weekly discussions grew in warmth and depth and became something special and formative for me. Having a kind of love affair with the oldest person I knew might have been carrying customer relationships too far … but it was not. It was a meeting of two people who seemingly could talk forever.

The door was open one Saturday morning when I turned up with Mrs Henderson's meat. I went in and heard a faint voice coming from the bedroom. She was lying covered up looking tired, old and sleepy. 'Hold my hand,' she said, and I did for a few minutes before she sent me away. The next day she passed away in her sleep. My parents could not understand my grief and I could not explain it. They thought I must be coming down with something. Nobody realised what I had lost.

In the meantime, the customer message had finally got through. Capacity and willingness to pay, it turned out, were more important than intellect. In fact, intellectuals were less likely to pay on time; they had other things to do with their money. The next debate was prompted by the pie-mince

contract. The local baker with a flourishing trade in pies had been reading
business books. Jim reckoned he had got as far as the Supply Costs chapter.
He had hit upon the pricing of the mince as a way to improve his margin,
and asked us for price reductions. We made an adjustment, which he
described as insufficient. The subsequent discussion focused on quality, my
father declaring he would not change his specification and reduce the price
by substituting fat for beef. He told the baker this would ruin his trade
and our reputation for quality. We lost the contract. Within a month the
pie was under attack. It was greasy and inedible. Within two months the
baker was back; the contract was ours, and at a higher price. The quality
message had been sent, received and paid for.

As we were learning the lessons of customer relationships and product
quality, so tutoring on negotiation began. There was no better time for this
than turkey-buying time. We journeyed to Aberdeen on a Sunday to buy
from an old and experienced farmer. We went equipped with marketplace
knowledge, but the condition of the birds was what counted. The farm
was large and the turkeys numerous. It took me a couple of hours to look
at them and their environment. Both, father pronounced, were good;
these birds are worth buying. This was the easy bit. After preliminaries
we talked about the surplus of birds on the market. Having bought, we
would have to sell within a limited time: there were no post-Christmas
sales for turkeys. And so it went on. We put numbers on the table and
haggling began. He brought out the whisky but there were still bridges to
cross. At last we arrived at an agreement and my father produced the cash.
The old man then recognized my presence and said, 'Come with me—I'll
show you where to put this,' and into the bedroom we went. He unlocked
a trunk and stashed the money inside. Pushing it firmly under the bed he
pointed to the shotgun resting on the wardrobe. 'There's the security,' he
said. Amazed, I left for the long journey home wondering if he would stay
safe with all that money under the bed, and whether he would still be there
next week when we went to pick up the turkeys. Luckily he was—and so
was the money!

*   *   *   *   *

Before I left my first school I had lost one hand, ripped a large hole in my
left leg and knocked myself out in an exuberant game of Relief (a mixture
of hide-and-seek and British Bulldog). The leg, gashed on a nail in a fence
while retrieving a football, had prompted the doctor to observe that I

could not continue to be so careless with my limbs and expect to reach adulthood. He had arrived on the scene that day in the nick of time; the football coach, armed only with embrocation a large bottle of iodine, was on the point of emptying the latter into the deep wound.

Towards the end of term, I was selected to address the school on the headmistress's retirement. This created mayhem at home. The speech was to be my own, the school insisted, but my father and brother interpreted this 'own' as 'from the Reid house'. They would be the designers of the text. By this time Jim was deep into Dickens and his language was not just flowery, not even floral, it was luxuriant in foliage and verbiage. His first draft ran to seven pages. No chance for an audience of under-tens to clap or laugh—just to stand, clustered around a small platform in a cold playground for a good 20 minutes listening to unintelligible gibberish.

My father then took his turn, starting with quotations from Burns and proceeding through strange jokes about headmasters to a thundering finish with a reading from John Ruskin. Mark you, it had the merit of being five, rather than seven, pages, and at least avoided mention of the misdemeanours of Simon Tappertit and the fortunes of Pip.

It was my mother who eventually sat down with me and said, 'What do you want to say?' We worked out three simple messages. We liked our headmistress because she knew us all and would always help us. Cheer One. We hoped she would come back and see us. Cheer Two. And she had made many friends from the time she had been at the school. Cheer Three and applause. We rehearsed it many times. We put in a laugh at the beginning and it was a great success. It is a formula much repeated since. Settle them down, get their attention, give the simple message and let them cheer. Listening to Jomo Kenyatta in Swahili 20 years later, I realised it works just as well in other cultures and other places, too.

*   *   *   *   *

The eleven-plus, despite all the adverse commentary, was no real hassle for a reasonably intelligent child. What followed was much more divisive and dangerous. We arrived at our new school in September to be corralled into a huge hall and then divided into groups labelled from A to HK. This cattle-penning activity branded inferiority on day one. If you were not an A you were destined for a subordinate role. If you were left for the D or the H you were a reject. No wonder discipline and compliance became problematic. I am not an advocate of comprehensive education in a form that holds back intelligent children, but I do believe the presentation of the package must be such as to stimulate involvement and full student participation. It must accord respect and it must uncover and develop the potential that exists in everyone.

My entry to an A class was a relief more to Jim than it was to me, since he had been extolling my potential to his friends, and any lesser classification would have meant loss of face. By this time he had already had one essay published in the school magazine. Ten others had been rejected in favour of inferior works for reasons, he asserted, ranging from wealth, beauty and religious prominence (the author being a minister's son). He worried that his brilliance, although erratic, might in some way be held against me. He had a nightmare vision of the headmaster sifting through the allocation lists and casting straight into the HKs any Reid who might be related to the Writer of Pamphlets and Scripts.

It was at about this time he began to tell me that I was adopted. He warned me not to tell my mother that I knew this, since she would simply deny it. But, he said, in certain circumstances I should not shrink from revealing it to the authorities. I think this was his insurance plan. If the simple fact of my being related to him endangered my grade allocation, then I should be prepared to let it slip that there were no actual blood ties, which would send the relieved headmaster back to the list to reinstate Reid minor (alias Rankin) to his merited A status. The absence of a Christening certificate—which I am sure Jim carefully withdrew for a time—lent credence to this tale. By the time the appellation A was read out, the pressure was off—but I still lay in bed at times and wondered: was I, or wasn't I?

Getting to know my classmates over the next six years was a great joy. The standard of teaching was high and this was borne out by the number of scholarships and awards the school took at the annual university competitions. In a way we were fortunate because several of our teachers had graduated in the 1930s when there was no money available for them to go on to post-graduate work, so they had taken to teaching.

A prime example of this was the classics master, Tom Howie. A poet, an idealist and a wonderful teacher with a lively sense of humour and great sympathy for the man in the street, he was a huge influence in my life. In our sixth year, the scholarship year, you could choose a broadening subject, which was dealt with on a tutorial basis. At his instigation I chose Greek, and for my thesis 'The Classical Influence on English Literature'. An impossibly broad subject for a sixth-former, but a wonderful entry point to the widest range of literature and poetry. It was a watershed experience. Tom's genius and enthusiasm unfolded treasures I never knew existed, created excitement and emotions I did not know I had. That brilliant summer we read the poems of A. E. Housman, one of his favourites. Particularly poignant was 'When I Was One-and-Twenty'. As the love-affairs of a co-educational school waxed and waned, Housman (despite his detractors) made good reading.

But beyond *A Shropshire Lad* we read the poets of the First World War and spent a long time on Shelley. As the year went past, Tom and I became firm friends and this was a new experience for me. A teacher could be a friend, could take you home to his wife and his unruly son and argue politics with you at length. The talk of different lands and new orders, of fairness in education, of the importance of free speech, was exciting stuff. When we played golf together with my father and brother and then retired to the tavern to argue and debate, it seemed as if a new world was beckoning. To open a book because you wanted to, not because you had to, was not a new experience for Jim, but for me it was a beginning.

Tom had always ignored preparation for the scholarship exams. He believed in cultivating the field before harvesting. His cultivation was a loosening of preconceptions, a removal of fear; it was encouragement to think and to express your thoughts. This process maintained both freedom and intellectual rigour, and these disciplines have served me well through business life. Ultimately you must form a view and you must act on it.

His last advice to me before the exams was to give my imagination full rein, to let it flow, to mark my application with personality and originality. This rang in my ears as I sat down in the huge hall. I did as he said and let it run. The sparks he had put into me poured onto the paper and I enjoyed it. I communicated it in my general paper, in my history answers and even in my Latin prose. Though I felt distinctly middle-of-the-road among the other articulate competitors, I was happy with my answers, and once the exam was over I went off to do what I did best: play golf.

Six weeks later I was on the tenth hole of the Eden course when my father appeared with tears in his eyes. I thought somebody had died. He shook my hand and said 'You have won the ninth bursary!' Tom Howie, a genius teacher and my second adult friend, had raised my game. He, his wife and the Reids celebrated long into that night.

By the time I left primary school, athletics was out and tennis was hopeless. It either rained and you could not get onto the courts, or you *could* get on the courts but the grown-ups pushed you off. Even if you wanted to practise, you needed an equally enthusiastic partner and usually there wasn't one. Golf, on the other hand, had everything tennis did not. You could practise endlessly on your own and bad weather was a plus: the golf course emptied. So golf it was.

Cupar, a nine-hole course considered the oldest in the country, had produced some fantastic players: Bob Peattie, who won the British Boys in 1924; Andrew Dowie, who had competed for top amateur position in the Scottish Open; and Charlie Todd, who won the Scottish Open Amateur. For some inexplicable reason this course, on the side of the hill eaten into by a cemetery, was an excellent nursery for the young golfer.

This is where I started, with plenty of help and advice. By the age of 13 I was slipping out of the house at six o'clock in the morning when everyone was asleep to practise for a couple of hours. It was both maddening and rewarding. Bad shot followed good shot. The swing got faster and the hooks became greater—tears of frustration flowed. But slowly it came. My father found a coach, Dave Ayton, who had worked in America as a professional and returned in retirement to St Andrews. His coaching was methodical: timing, tempo, and no temper. Start with the short shots and work up to the long. 'Show four knuckles, left heel down—control the swing and see the shot ....' These words echoed in my mind and eventually I began to win some matches.

The British Boys Championship was to be held in St Andrews when I was 15. My brother was going to enter but my father forbade me to until my handicap had come down to single figures. When it did he still demurred, but after consultation he relented. The intensity of practice grew until the day came. The local papers had a story and interest in my golf grew, more from lack of other news than anything remarkable.

The first tee was nerve-wracking and I lost it, and then the second, the third, the fourth, fifth and sixth. The steward on the hill on the seventh, some university worthy, said 'Simply stick in,' and at last I got a par four and a half. A two at the eighth brought one back; an opponent's lost ball and a four on the ninth, another. On the tenth my ball ran through the green and I was faced with a 25-yard run shot that Dave had made me practise again and again. I hit it well—I had to get a half—but it ran up the hill straight as a die into the hole. Three back, and on the 11th, which had a wicked sloping green and cavernous bunkers on either side, Dave had advised to me to play short. 'Line it up,' he said, 'and with a good chip you have a three.' I played short but my opponent played a magnificent shot four feet behind the hole. In those days it was stymies: if one golfer's ball sat on the green between the hole and the opponent's ball, then it stayed there. You had to either chip over it or putt around it. Mine ran over the hole and ended two inches past it, directly in front of his ball. He putted and his ball ran off the green: one more back. As we passed the steward, he dropped his flag and began to follow the game. In St Andrews people emerge from nowhere when a game is on, so the crowd grew. The run continued—a good seven-iron shot eased me onto the 12th green to within five feet, and it was another three. The 13th was not good and it was back to two. But on the 14th he had three balls out of bounds and it was one again. With a lovely four on the 16th it was square, and there I was on the 18th with a six-footer to win on a gently sloping green. It slid smoothly into the hole and the huge crowd erupted. I was totally dazed. What had happened? Where had all these people come from, and how had I possibly won?

We escaped from the reporters and went for a long walk along the beach, trying to take in what had happened. The next day it was another teenager's turn to draw the crowds, and I was beaten.

It could not happen again, but five years later in the final of a St Andrews club championship (the Jock Hutchinson Cup), I finished on the losing side on the last green having been five down with six to play.

These years at St Andrews learning the game deepened my love of it. Golf is a contest with yourself. It tests your ability to keep your game together—to rescue something from your mistakes—to find a wave of success and ride on it. Just like business, as I was to find out. Jim prophetically said, golf is 'not over till the fat lady sings'. Out of curiosity I traced the aphorism—it gained public acclaim when a baseball commentator used it in a famous game, but its origins lay in the Deep South, where a mother told her son that they could not leave the church until the fat lady sang. Of all my golfing experiences and exploits, none compares with that day when the fat lady sang on the last green at St Andrews.

While Cupar prided itself in its young membership, the young did not always behave well or respectfully, particularly towards the at-times-officious committee. My brother was always at the forefront in these spats. There was some contention regarding starting times—juniors could not start after five o'clock in the afternoon. Jim preferred to engage in fruitless debate about players' rights, while we younger players had a more pragmatic approach. We simply adjusted the club clock to give us an extra half-hour's play.

When The Open came to St Andrews after the War, the members
flocked in. We played truant and went to watch Sam Snead, an American
serviceman, win the claret jug. The committee was amazed and delighted to
see the note from him in the visitors' book thanking the club for letting him
practise early in the morning. The entry won huge publicity, but the effect
was short-lived when a new note from Henry Cotton (who won The Open
in 1934, '37 and '48) appeared, castigating the course. He complained
that his having to practise on the side of a hill alongside the cemetery with
a mad American hitting balls all over the place had ruined his chances.
This was too much. There was an internal investigation. Henry Cotton's
signature was valid—or it certainly *looked* valid, being very like the one in
his new book, a copy of which my brother had received for Christmas. No
progress was made. The juniors were warned. Pointed looks stopped short
of accusations ... and the visitors' book was withdrawn.

                    *   *   *   *   *

University was a change of pace. Gone were imposed discipline, set
timetables and homework. In the first year, Latin was easy and medieval
history well taught. The general exams presented no difficulty. I was
interested and involved in student life, and golf filled my spare time. By
the third term, selection for Kate Kennedy was under way. In this, one
of Scotland's oldest historical pageants, students and citizens gather to
watch a colourful procession leave St Salvator's College gate. The original
Kate was the niece of the college's founder, Bishop Kennedy, and in the
procession her part is traditionally played by a first-year male student
whose identity is a well-guarded secret until the last moment.

   Selection involved a series of interrogations that just seemed fun
for me as I had already ruled myself out as a candidate. A one-handed
Kate Kennedy would defy all historic tradition. But the senior students
thought otherwise, and I was elected. This was an honour I did not
expect and it led to my being pitchforked into the Students' Union
committee, and then almost immediately into the role of treasurer. The
Union was effectively a large business with no administrative staff; our
practices of cash disbursements for entertainment pretty much resembled
those of a Wild West saloon. But gradually, by changing key committee
members and shifting responsibilities, we forged some order from chaos
and the revenue gains began to provide financial stability as well as
reserves. The process was time-consuming and distracted from academic

study, but it was enormously valuable as an introduction to business management.

In that second year of union office I established my credentials for the vice-presidency and put in place a treasurer who would sustain and carry forward the reforms we had made. The union was a robust place, at times too robust, but peace could always be restored.

It was in the second term of that second year that disaster struck our group of happy students. A fellow medieval history student, Sheila, and a medical student were found dead together in his flat. Rumours were rife, but whatever had happened, the loss was appalling. For a while the university became a quiet place. Men and women hugged each other. Some female students who had felt the arrogance and cockiness of their male counterparts in daily exchanges were quick to apportion blame, but they had no knowledge of the actual situation. The deaths cast a shadow on the term and deep down we wondered if it could have happened to us. I had attended a lecture that day with Sheila, a communicative, happy person … but one, it turned out, who was destined for death within 12 hours. To this day she is often in my thoughts, her friendly young face and bustling red-gowned figure hurrying by.

*   *   *   *   *

Miss Ketelby made modern history exhilarating; looking for the forces at work in a political situation and identifying potential outcomes— then contrasting those with what actually happened. This would have continuing relevance to my future life, particularly in the convulsions of Nigeria post-independence, in the Vietnam War and the Khmer Rouge, not to mention Iran. The American War of Independence and the French Revolution had to do in the interim. Economics became more interesting when we moved on to trade unionism and banking and money.

In philosophy, which was compulsory in the MA curriculum, we had to take a course in logic and metaphysics. This was taught by a tortured professor who found 'being' difficult to define. As he was a regular visitor to the Royal & Ancient Golf Club (but not for golf and not much for drink either), I knew him, but he would not acknowledge me. Perhaps to recognize one student would raise the question of how he knew the other 1,200 existed when they were not actually in front of him.

He set an essay on Descartes and *cogito ergo sum*, 'I think therefore I am'. I was challenged by this and worked long and late on it. I contended in my essay that Descartes had it the wrong way around, and perhaps this was a mistake not only in presentation but also in substance. It should have been, I argued, 'I am, therefore I think'. Am-ness was the key. Am-ness was what directed thought. Thinking was a process ignored by some because of their particular-ness, their blind absorption with self. Others' am-ness was found in the individual creativity that wrote books, painted pictures and made music. I built on this thesis, submitted it and awaited the return of the paper.

When the professor gave back the essays, mine did not appear in the first grouping, or the second, third or fourth, and I wondered if I was going to get it back at all. At last he said, 'I have a paper here which I have puzzled over. It could be a work of rare distinction but that is unlikely. It has thought—individual, disturbing and radical thought—but it doesn't touch the subject. I have therefore after much thought of my own given it a mark of distinction: zero.' He added, 'I could be wrong,' putting his hand to his head apparently in despairing anguish. This aroused my classmates' interest in the essay, and they stamped their feet in my defence, but the mark remained unchanged. That evening I read the essay that had come top, and it seemed to me to be simple regurgitation without a single original idea. If that was what he wanted, I thought, I would deliver, and I passed easily—but with very little satisfaction.

After a year as vice-president I was nominated for union president. I agreed to stand after discussing it with my friend the secretary, who had declined to stand. There were no other candidates until one minute before closing, when the secretary allowed his name to go forward. His apology was hollow but accepted. I told him I would lose as, having moved out of residence, I had no political base on which to fight, and so it was. Defeat can be a good thing if you learn from it. My first mistake had been to trust a friend without recognizing the strength of his ambition. I was also beginning to recognize the role of 'king-makers'—people who will manipulate to influence power without themselves being candidates for the role in question. Any situation where conflicting ambitions are at work is a field for the king-maker.

I did, however, become president of the Kate Kennedy Club, and this supplied further experience and preparation for corporate life. My administrative aptitude had improved significantly while I had been treasurer, but it still left much to be desired. By contrast the club secretary, Henry Sefton, had an exceptional talent for tidiness and organization. An austere divinity student who subsequently became a senior lecturer at Aberdeen University, his minute-taking and follow-up agendas were far beyond anything I had ever seen. He produced smoothness and logic in what were truly complicated arrangements for a nationally acclaimed pageant.

Meanwhile, golf had become a major part of my life. When the captain opined that, having only one hand, I was unlikely to withstand wintry weather, my close friend Dave Nisbet said, 'Why don't you just beat them all?' So one by one I worked my way through the team until I reached the captain, who at that point dropped his objection. Dave and I began a successful partnership which was to last for four years. In our final year we lost to only one pair—a pair that included a Walker Cup player—on the last green. And though my singles record was not bad, thanks to Dave, an incredible putter, it bore no comparison to our foursomes success. Dave wore thick glasses and was always vastly underestimated, but his snooker was superb, his darts fantastic, and his golfing prowess confirmed in later years when he qualified for The Open.

In our third year we passed on to advanced economics. As I sat beside a good friend in the first class of term, I noticed an animated redhead dressed in a strange half-coat. But it was neither the mass of curls nor the coat that attracted me; it was the vivacity with which she captivated her neighbours. 'Who is that?' I asked my friend. He replied

'That is bad news, Bob—stay away!'

And so of course I did not.

Joan Mary Oram lived with her cousin Christo, a Scottish golfer and sister of the Queen's surgeon, Professor Donald Douglas. Christo welcomed me with open arms and monopolized the conversation on golf, politics and much else. It was refreshing and created a warm and stimulating environment in which Jo and my relationship blossomed.

The remaining two years passed quickly. Jo and I met every day in the tea shop and discussed economics and history endlessly. Her views were well researched and her interests encyclopaedic. She was everything that, nine years earlier, my childhood friend old Mrs Henderson had wished for me.

*   *   *   *   *

The exams came and went, and (unlike me) Jo won first-class honours. My job-hunting had yielded offers from ICI, Unilever and Shell. I liked Shell best of all. I liked the people. I liked the excitement of foreign exploration, and oil seemed to be a growing business. The snag was that the date for the final interview was set at nine o'clock on the morning after my parents' silver wedding anniversary party. This turned out to be a riotous affair and I did not get to bed till well past midnight. The following desperately cold morning I left at seven o'clock to drive 40 miles to St Andrews in a jeep. When I arrived, half-frozen, the room was warm and the welcome genuine. As we sat down beside the roaring fire, the telephone rang. Like any good Shell man he asked permission to answer, and then was drawn into what was clearly not going to be a short conversation. The next thing I knew he was gently shaking my shoulder to wake me. I proffered many apologies but he dismissed them. Assuming I had blown my chances with the interview proper, I relaxed and we chatted for over an hour, at the end of which he promised he would be in touch.

The next day my professor asked me to stay behind. I assumed I would be in trouble for my somnolent performance. 'Well done,' he said. 'The job with Shell is yours.' I was overcome and stuttered my thanks. 'I have a question,' he said. 'Your interviewer told me that in all his years as a recruiter he had never met anyone so relaxed and so open. Why could this be?' It was on the tip of my tongue to say, 'I always am,' but I settled for: 'I've no idea!'

The dates were fixed. I was to spend three months in London being inducted, and then be posted to Borneo. By now I was ready to look life in the face and hoped to persuade someone to come with me. I had no doubt that it was Jo I wanted to spend the rest of my life with, and I told her so. She said if I promised never to betray her she would marry me and come with me and share my adventures. I promised, and I kept my promise. After I had sought and gained her father's permission in the traditional way, she and I set off to accumulate funds for an engagement ring.

We found jobs in a hotel in Gatehouse of Fleet: Jo as assistant cook, me in the bar (where I distinguished myself by heating an empty coffee pot till it exploded) and delivering milk. The early-morning milk run was around a caravan site. Most of my customers were still asleep, but one middle-aged man was always sitting on his step reading a book. He thanked me for the milk and began to inquire where I came from and what was I going to do. By the second day he knew I was about to join Shell and was going to Borneo. He suggested I stop by for a cup of coffee the next day so he could tell me about Borneo, so for the next five days I was treated to a personal seminar on Sir James Brooke, the rajah of Sarawak; the Brunei pirates, vivid pictures of Kuching, Miri, Brunei and the rivers that were the highways into the *ulu* (jungle). He urged me to learn Malay—which I later did from a frozen but friendly Malay, both of us hunched in front of an electric fire in a London bedsit. Caravan Man quietly disappeared with his store of knowledge, but by then he had fired my imagination about Borneo, its culture and its people.

* * * * *

Once Jo and I had saved enough, we bought the ring and were set. Jo would lecture for a year or even two while I went off to my first posting, and I would set about convincing Shell to let her come out and join me.

2

Borneo

The flight to Singapore should have been uneventful, but I had never been on a large plane before, so it was exciting. My neighbour claimed to be an experienced flyer but was wildly anxious about the Suez crisis and resulting flight diversions. When the approach to Basrah was aborted, his anxiety went off the scale. Drawing on my experience of movies, in which planes regularly abandoned approaches and zoomed away to try again, I calmed him, saying this was entirely normal on diversions. And so it was: we landed safely second time round. He would not leave the aircraft, preferring to stay on board with his gin while I went to take my first look at the Middle East. As I left the cool of the plane I was hit by a wall of heat and blinding sun. This was new. The terminal building was not. Air conditioning, which I had heard about, was obviously not installed and the half-a-dozen fans lazily circulating the hot air had little effect.

Back on the plane, my neighbour was inquisitive about the terminal so I reassured him about its facilities, and particularly the fans. He was not impressed, and stayed that way until he left me in Bombay. When we arrived in Singapore I discovered that you do not supplement air conditioning by opening windows. I obviously had a lot to learn.

We left Singapore early in the morning, driving through the streets just as the houses and the shops were coming alive. Our destination was Labuan, a tropical island off the coast of Brunei with beautiful beaches, palm trees and vegetable gardens cut out of the jungle. The rest-house where we spent the night overlooked the beach, and from there you could see the clear blue sea—so enticing, until you saw huge stingrays waiting in the shallows. Could this encapsulate the East, sublime beauty on the edge of danger? It had certainly been so in 1945.

The taking of Labuan was a critical part of the plan to capture Brunei Bay and secure a base for the offensive on Brunei and Sarawak. Australian forces met aggressive resistance from the outnumbered Japanese, but after heavy naval bombardment they eventually submitted. About 400 Japanese were killed and 11 captured. The Australians lost 34 soldiers. I was to learn more about this from one of their comrades: my first boss. The war graves and the memorial brought harsh reality to this paradise island, and a sharp reminder of a war that had engulfed not just mainland Europe, but the world.

We joined our connecting plane and flew to Anduki's grass-strip airport in Seria. The descent took us over the mudbanks of the Belait River, where crocodiles rested in the sun. Stepping out of the plane into glaring sunlight in the middle of the jungle was different again. Different from St Andrews, where the sun never glares and the woods are orderly, not tangled with the creepers and low bushes that dominate the Brunei jungle. Different in that it was all utterly new and strange.

From there we headed off across small rivers edged with mangrove trees and black mud riddled with crab-holes. Gas-plant flares and workshops gave the first indications that we were entering an active industrial complex. Beyond them, past the busy marketplace and back to the seashore, you could see a platform at work, way out at sea.

A narrow pathway led to the bachelor quarters. Our baggage, dropped unceremoniously at the front door, was picked up by the Chinese cook, who took us to our rooms. My friend Don and I were given the inside rooms since the outer rooms, cooled by the sea breeze, were already occupied. The bedroom was basic: a bed and a mosquito net, a cupboard and a chest of drawers. A lavatory and shower completed the set-up.

Having taken stock of our position we decided we would have a beer. In the fridge, each bottle was marked and it was obvious that none was ours. This did not deter us; we reasoned we could replace any we drank—but when our fellow residents returned they did not share our view. Don's impressive proportions and the trust they put in a Scotsman's word helped us past this moment of tension.

Within a day, one of our two housemates returned to the rig for three weeks. He had had his week off. The other, Jan, never went out, and we seldom stayed in, so breakfast was our only point of congress. After a while the ice thawed and Jan, a Dutchman, began to talk to us. Forty-odd years old, he had been around oil camps in Venezuela and Indonesia since the war. His career was solid but by no stretch of the imagination could it be called galactic. In fact as far as I could gather he was bored out of his mind. Apart from an occasional outburst about the female sex—or rather the absence of it—in the camp, he seemed to spend his time silently in his room either studying or reading. Our quarters were elevated five or six feet off the ground, balanced on sturdy concrete posts, and at one point he toyed with the idea of constructing a trap door through which he could introduce lady guests ... but later abandoned the notion as impractical.

As Don and I continued our induction programme, dined, golfed, drank and played tennis to fill our days, we thought less and less about Jan. He just seemed to watch wistfully as we rushed out to the next party or the next game. Then one day I heard him arguing with Don in the sitting room. I joined them. Jan seemed quite upset, but after a few minutes he turned to me and said, 'You would do it, wouldn't you?'

'What?' seemed a reasonable question. The answer was extraordinary.

'Let me predict your life story.' This seemed to be no big deal. After all he could not do it and even if he did I would not believe it. As we sat there—him wide-eyed and distraught, Don disturbed by his weird reaction and me nonplussed by the absurdity of the situation—the cook put supper on the table.

At the time, I was reading Somerset Maugham's *Far Eastern Tales* and was disappointed that I had not yet come across anything that was in the least extraordinary. Here on my own doorstep was the making of a good yarn, so in the end I gave him the date, time and place of my birth and even the name of the by-then-deceased midwife. Don was relieved in one way but appalled in another.

At first it seemed to have worked. We hardly saw Jan for weeks. When we got home from work he was in his room. When we went out, dinner was set for one. When we returned, either together or separately, the light was on in his bedroom, and he even took his breakfast in there, alone.

Six weeks passed before he caught me one evening and said that it was done. I felt like Dr Faustus as the moment finally arrived. The next day we stood together over a huge chart covered with stars and lines that all seemed to move together in a hallucinatory way. Melodramatically he pointed to the junction of one heavenly body with another. Tracing the point of collision with his finger, he cried, 'This is where it all began!' He was fiercely intense and barely in control. In as normal a voice as I could manage, I said, 'So that's when the gods delivered me up?' He spun around and collapsed into the chair.

'Think!' He cried. 'August 1943!' It dawned on me that that was when I lost my hand. 'This was the battleground—this was the field of wounds. This was your great escape!' he exclaimed.

I called for tea and played for time. I suggested he look further back and this calmed him. Tea arrived. I said, 'Jan, you have frightened me with your vision. I know what you're telling me and you're right, but I can't contemplate knowing about the future. If I did, how could I live my life with any conviction?'

Crushed, he replied, 'But you will love someone, and knowing that will help you—I can see that clearly.' I told him firmly I did not want to know any more. He began hastily to gather up his charts, and when Don then came back from tennis he urged me to get away. Jan fled to his room, slamming the door.

The party that evening was no fun for me. The business with Jan had disturbed me and I felt stupid to have got involved. I also felt I owed him something and promised myself I'd make it up to him. Perhaps a river trip, perhaps a party—perhaps an excursion to Brunei town … I would work something out. When we returned after midnight, the house lights were ablaze—our cook and the cooks from the other bungalows were perched around the wooden steps like a flock of starlings looking for crumbs. Jan had been taken away, with all his clothes, in a white car. Did I think he would be back for breakfast? I shook my head.

His bedroom door hung open, half off its hinges, and inside, every square inch of the walls was covered in charts. These were his calculations about my life's trajectory, my past, my future; a story he thought I should want to know, and could not understand why I did not. I was not allowed to see him again. It would have been bad for him and not good for me. 'Repatriation' was the word they used, and the charts were packed off with his other belongings. The camp had it that he was of royal blood, too pure-bred for the oilfield, too sophisticated for his work, too intelligent for his colleagues. Just another casualty to be chalked up. It was a sobering and sad experience.

*   *   *   *   *

The house we lived in had been shot up by the Australians as they chased the Japanese, and since we tenants were just single men, nobody had bothered to repair them. Mosquito nets smelling of antiseptic filtered the light. Getting in and out of the building was a dangerous kung fu operation. Our cook, Pussy, a small muscular man of indeterminate age, guarded all access points ferociously. 'Hit before ask' was his motto. It took a while to get used to being woken each morning by this barefoot warrior prodding you with a broom.

Oil fields are dangerous places and we were living and working right in the middle of one. Within the first three months a gas well blew out, shooting stones high into the air and leaving a cloud of gas hovering over the well. Tension grew as the hours passed and the cloud sat there in the still heat, refusing to disperse. Attempts to cap the well failed, the crane foundering uselessly in the muddy lake created by the fire hoses. At last a team got close enough to guide the blow-out preventer over the well, and then slowly close the valves to kill the jet of gas. It was, the experts said, a shallow well on a shallow structure, so shallow that the gas had blown its way along the structure and come out under a large concrete batching plant close to a swamp. The plant had disappeared into a sea of mud, never to be seen again—except, perhaps, as an underwater playground by the local crocodiles.

If this had happened today, the engineers would have been mired in a succession of royal commission reports, legal cases and calls for resignations. Instead, we got on with it and used what we had learned to operate and develop more safely. We knew we had been lucky. If an onshore wind had wafted the gas over the flares and it had imploded, there would have been a conflagration to match anything the island had seen in the Second World War.

This major event put into perspective the market riots and the stabbings, news of which punctuated my days in my first job at the labour office. The riots seemed often to stem from drunkenness and competition for the favours of the female sex, who at times—surprisingly—turned out to be male. The he/she was not a gender then much evident in Scotland, and certainly not in Cupar. They were beautiful boys with apparently curvaceous figures dressed in sarongs and colourful blouses. They wandered through the camp enticing the unwary until all was revealed— then suddenly the pussycat became a tiger, scratching and biting until money changed hands. The process much amused those who had seen it all before.

My first market riot in Seria was not about drink or sex as everybody assumed. It was about money. Some Hong Kong Chinese technicians had run up huge gambling debts which they could not pay. They decided they had to get out—fast—so they set up a riot which, amid the frenzied excitement of Seria nightlife, was not too difficult. They managed to get themselves arrested and charged, and were sentenced and returned to Hong Kong, thus escaping their debts and their outraged creditors. Had the police rung their opposite numbers in Hong Kong, their escape might not have been so easy.

There was much to learn in these initial six months. Barney, my boss, was an able teacher although he was looked down on by some of my well-educated colleagues. He had in spades what many of them would never have, though: he was acutely perceptive. He could see right through pretence and posturing. He had landed in Brunei as part of the Allies' liberation mission in 1945, a sergeant in the Australian infantry. Some of the Japanese withdrew to an inland stronghold, forcing Barney's company to search through dense jungle and muddy swamps. He found and had to deal with captured soldiers on whom the islanders were taking violent revenge for the atrocities they had suffered at their invaders' hands.

Barney's army experience equipped him well to deal with stabbings. He simply removed the knife from the assailant, either by persuasion or by force. This job I had was like being in an X-rated movie, and when we moved in on a pornographic ring I began to wonder whether I had swapped my career in industrial relations for one in the vice squad. The ring catered for sex-starved workers who wanted neither boys nor the pox.

When the case came to trial, it transpired that the Indian community had hired David Marshall, a Eurasian former chief minister of Singapore, to defend their accused compatriot. David's name attracted crowds to court in the expectation of verbal fireworks and drama. They got it, but not in the way they expected. The elegantly attired Marshall listened carefully to the charge and, his face a picture of increasing repugnance, to the long and lurid list of evidence.

He then rose, repeated the charge—of being in possession of pornographic materials—and called for the case to be dismissed. Possession, he explained, though distasteful, was not a criminal offence. The crime lay in the purveying of these materials, and that was not the charge. The judge had no alternative but to drop the case. Marshall asked to be allowed to burn the materials in the court's own yard, so the circus ended with a bonfire, ceremoniously ignited by the defence, and the discomfited departure of the prosecution. Barney shook his head but grudgingly had to admit that Marshall was a class act.

Our managing director, Hector Hales, seemed a remote figure to me, a new recruit on this large and busy camp, so I was astonished when his PA informed me that I was invited to supper at his home. His *kajang* house was made of *atap*—woven palm leaves—and floored with teak beams. It was furnished in a relaxed way and, away from the office, Hector was clearly relaxed, too. The supper party was in honour of his weekend house-guest. When the guest appeared, I was amazed to see it was my caravan-milk-round customer, my personal Borneo tutor from before I had left Scotland, who turned out to be the governor of Sarawak, Sir Anthony Abel.

Late in the evening, the conversation turned to oil and to nature—and to eggs. Hector maintained that you could throw an egg over the roof and it would land unbroken on the lawn. Dispatched with half a dozen eggs, we duly threw them over the roof and four landed unbroken. We suspected that the ones that landed intact must be hard-boiled, but empirical research proved otherwise. We then had a physics lecture on the egg, and from there a lecture on gas and oil structures.

Hector had decided that it would be a good idea for his young trainees to see the real world outside the environs of the camp, and Sir Anthony agreed. The logic was sound. The future success of oil investment rests not with the operator, Hector explained, but with the owner—the country of origin. To do business successfully requires a knowledge and understanding of the host nation's priorities and aspirations.

He was thinking of sending a couple of his trainees to Sarawak to learn how the colonial system worked and how the state was progressing to self-government. He realised that the oil industry would have to adapt if it was going to survive in a changing environment. It was an exciting discussion because it was not philosophical or theoretical; it was from the front line. It exemplified the essence of Shell's success from the days of its founder, Marcus Samuel: decisions being made at the sharp end, not in a distant boardroom. It was an educational evening, and made that crack-of-dawn milk run in Gatehouse all the more worthwhile.

Shortly afterwards, to my surprise and delight I was instructed to pack up and go with Don to Kuching and to Simanggang (now known as Sri Aman). As we sat strapped in on a bucking DC3, it seemed that the aircraft's mission was to look for cloud and storms and then fly through them to test resilience. We certainly achieved the objective.

Hal Ferguson, a young assistant district officer who was to provide our base and be our guide, met us at the airport. Early the next morning we fought our way through the bustling city to the docks and onto the boat that would take us up the Batang Lupar to Simanggang. The boat was packed; people and goods vying for what little space there was. Facilities were minimal. The toilet was a platform protruding from the stern, not much privacy provided by a sacking curtain. Nearby passengers-turned-spectators judged and giggled at the quantity and condition of the output. I was reminded of this human fascination when I watched *The Madness of King George*, in which the contents of the royal bedpan are much discussed. A blue emission into the river, I feel, would have unbalanced our boat with excitement.

The Batang Lupar is famous for its bore. This huge wave carries all before it, sweeping up the side of the river, snatching anything in its way, overturning canoes and setting them adrift. We watched as it sped ahead of us, sending shrieking children running up the banks and canoers paddling madly for the shore. It was hot and becoming hotter and the jungle grew denser as we made our steady progress.

The bore figures in Somerset Maugham's 'The Yellow Streak' (a story in his book *The Casuarina Tree*), in which a Eurasian saves himself and leaves his lover to drown in a capsized canoe. This tale reverses the facts of an actual fatal incident in which a Eurasian died saving his lover. The misrepresentation apparently so incensed the officers of Sarawak that, after the story was published, Maugham was refused entry to Sarawak at the Port of Miri, and left irate and disconsolate on board his steamer. Strangely he never wrote of this rejection!

At around midday, a substantial town emerged from the haze. This, we decided, must be Simanggang. We were almost trampled underfoot by our fellow travellers, all anxious to be first off the boat in spite of their junior colonial master's exhortations. He, meanwhile, courteously handed children's baggage over the side of the boat onto an overcrowded quay.

We threaded through the crowd up to the house that was to be our home for the next three months. This one had not been shot up, either by the Japanese or by the Australians, but it had similarities. Its wooden planks had been soaked in kerosene to deter ants and woodworm, and mosquito nets hung over the beds. It was a comfortable bachelor pad. The veranda overlooked the town and in the distance, the river, and at the rear the jungle, which stretched hundreds of miles into and across Indonesia.

That night we were to have dinner with the resident district officer, Griffin, who was something of a legendary figure. In 1942, as the British government's representative, he had had the responsibility of surrendering and handing Kuching over to the Japanese, but not until he had led a group of evacuees through the jungle to the Indonesian border. It must have been hard to follow the call of duty and turn his back on his own city. At the time he did not know that the Japanese were rapidly gaining control of the island so the evacuees were only exchanging one POW camp for another. On his return to Kuching he discharged the necessary formalities before being incarcerated with the others.

Griffin took up the story. The prisoners' camp was stable; nobody was going anywhere as there wasn't anywhere to go—there was no River Kwai to march to. So he applied himself instead to improving the conditions there. He made a case for a religious service on Sundays, and permission was granted. The bishop duly arrived with his dog and staff, but left with just the staff. He made his way home dogless, and the inmates acquired a valuable source of protein.

After the cessation of hostilities, Griffin rejoined the administration in Kuching. It was then becoming clear that the Dyaks (the native people of Borneo) in the Second Division had not signed up for the ceasefire. They continued the hostile activities that had disrupted the invaders, making the district a no-go area.

The river that brought boats upstream ran deep on the jungle side but was too shallow in the centre for outboard motors, so boats had to hug the bank. This made the passengers easy prey for the head-hunters, who used blowpipes with pinpoint accuracy and followed up with razor-sharp *parangs* (machetes). With the journey of the day still vivid in my mind, I could picture the travellers' dread as they made their way through overhanging branches, peering nervously into the dense, dark jungle.

Unfortunately, once the enemy had gone, the former guerrillas switched their attentions to civilians, who could be lucrative targets. Administrators devised schemes to offer alternative income streams such as subsidising rubber and pepper plantations, but these received only lukewarm responses. The chance to work in the oilfields, however, was more attractive to young men. Griffin, knowing the historical context so well, put the case for it very convincingly. He described his vision of independence, which would sweep through the whole colonial empire. India was only a beginning. A fierce advocate of grass-roots education, he had incorporated into our programme a session with his newly created council to show us how it worked.

At the end of a long evening we left him, me with my marching orders to Lubok Antu, 'the pool of spirits', and Don to the rubber-tree plantations of the lower regions. We would meet again at Easter.

I set out at dawn. In my knapsack I had a medical box packed with antivenom, needles, salt tablets, diarrhoea capsules and every sort of bandage. We were to travel by *prahu*, a long boat with a powerful outboard motor, driven by one Dyak with another as bowman. Many Dyak men were tattooed across their chests, up their necks to the chin and down their forearms. Our impressive-looking crew moved silently, like large cats. They did not say much and neither did they react visibly, so you had no idea what they were thinking. I wondered whether this would change as we travelled together.

We stopped at Engkilili, a large village with an array of shops that served the pepper gardens run by the Chinese. Having left their own country to seek their fortunes in a quieter place, they had ended up in Engkilili and built a prosperous business there. Being isolated and such a closed community, they tended to be suspicious of change, so it was difficult to gauge what the new Sarawak would mean to them, but in Griffin's scheme of things they would certainly be involved.

Past Engkilili the river narrowed, and the further we went, the closer the jungle seemed. Progress was slow as the stream pushed against us. The helmsman steered an erratic course through the current to ensure we were not decapitated by overhanging branches. Eventually we came to a village where district officer Peter Tingom was on the bank to greet us. We were to stay the night at his house, which was perched on a rock high above the river.

A smallish man bearing the Dyak tattoo, Peter had clear, friendly eyes and a warm handshake. As we made our way to the house, he explained he had spent time in Australia on a government programme and had enjoyed it immensely. His job now, he said, was fascinating—but difficult. His objectives were clear, but implementing them was another matter.

His wife was waiting at the house to see what the boat had brought in. During supper they quizzed me on my background and what had brought me to the oilfields. They were intrigued to learn that we had begun to look offshore. What sort of a fishing boat would you use? they wondered. They spoke enthusiastically of the opportunities the fields held for Dyak communities, so a consistent theme was emerging. Peter outlined our itinerary. We would travel in the jungle, partly by prahu but mostly on foot. The terrain would be difficult but not impossible. Objective number one was to visit the longhouses (or 'river houses') to assess the adequacy of their rice stocks. 'We need to establish if the stores will last till the next harvest,' he said. 'The further we go, the worse the shortages will be.' He explained the traditional practice of shifting cultivation, which begins when the steep riverbank near a new longhouse is cleared and planted. Once the vegetation is gone, rains wash away the fertile topsoil, so after the harvest the field becomes a wall of mud. The farmers then move on, shifting the area of cultivation to fresh land further from the house to start the process again. After successive shifts, the fields are far from the house, the farmers are older, the children reluctant make the extended journeys ... so the system is on a downward cycle. In extreme cases, the government was trying to move the houses and their inhabitants to the coast, to stable farming areas where help was available to get them started. This was objective number two.

Moving longhouses and resettling their occupants seems a long way from building offshore platforms, but this was my lesson. To the provider of the lease, the owner of the oil—the people of the country I was living in—having enough to eat was infinitely more important than drilling speculative holes in the seabed.

I lay in bed that night thinking of a bitterly cold fortnight I had spent working in the Lomond Hills. During a Christmas vacation I had collected sugar beet from a half-frozen field to help a farmer who was sick with pneumonia. As I threw the beets into the trailer, hoping the tractor's brake would hold on the steep slope, I asked myself a question that Peter must have wrestled with: does it have to be like this? Change for the Scottish farmer would have meant abandonment of all he had struggled for, but what he did not see was that sugar beet was doomed anyway—before long it would be driven out of the market by sugar cane grown in warmer climes.

I dreamt of numb fingers and hard, white beets but woke at dawn with warmth flooding into my room. When we had arrived the night before, it had been too dark to have a proper look at our surroundings, but now from the veranda I could see Borneo stretching away to the horizon, covered in white cloud. The cloud melted away in the sunshine to reveal a gigantic forest, massive trees standing out above the all-encompassing jungle. It was the kind of dramatic, mystical dawn depicted in adventure stories of far-off lands, but there it was, spread before me. These were going to be two weeks I would never forget.

For the first part of the onward journey the water ran deep and fast. It would be great coming down, but it was hard work going up. After a couple of hours we reached a clearing where we pulled in and clambered up the steep muddy bank. This was an exercise in itself, though the drivers made light of it. Peter struggled a bit—I was relieved there was at least one other amateur in the group. Once the prahu was secured, we set off along a well-trodden path. This was jungle walking: slippery, densely overhung on both sides and no idea what lurked within the green, entangled mass. Birds and moths flew up as we passed. After about an hour the ground became clearer and we could see gardens and small fields. At last I caught my first glimpse of a longhouse.

This was a very long longhouse, with rattan walls and a thatched roof, and it was where I was to spend the night. It was raised on stout tree-trunks and underneath were pens for pigs and hens. Children, mothers carrying babies, elders and young men (not impassive but smiling) came out to welcome us, and giggling young women naked from the waist up hid behind them. Peter formally introduced himself and explained who I was. He spoke mostly in Dyak, which I did not understand, but some in Malay, so I got the drift.

We were ushered into the house, which meant climbing a large upended log notched with a set of rough steps. Leaning forward you had some momentum, but coming down without tumbling headlong was going to be a challenge. A longhouse has a thoroughfare used by everyone, on one side of which are family homes and on the other open spaces for each household. Fighting cocks were tethered outside, one for each of the 46 rooms. One greeted the early dawn with a shrill alarm, prompting 45 more to follow. When they were all in full song, the cacophony was deafening.

Visitor entertainment was deferred until Peter had held talks to explain the new council arrangements. The young men were enthusiastic but the elders were more cautious, keen to ensure that their own power base would not be eroded. The discussion then turned to the oilfields and it was my turn to speak about the company, its achievements and its plans. I surprised myself with my confident presentation. As far as my audience knew, I was the managing director—and that you could get a taste for.

When darkness began creeping in, we turned to drinking. The choice was a white rice beer or arak, a fiery spirit that warmed you to the pit of your stomach. Go slow was the advice, and this required skill. As the spirit flowed, the dancing began: war dances performed with parangs, which glistened in the candlelight and threw ghostly shadows on the wall. At long last I slipped under my mosquito net and fell asleep, the security of youth leaving me untroubled by the human skulls that hung above me from the ceiling.

The next day I woke most pleasantly with three Dayak maidens stroking the blond hair on my arms. They ran away laughing as soon as I opened my eyes, but it started the day well and deferred at least my concern about the log-stair of terror. As it happened, with a hand resting on the shoulder of the man in front, there was nothing to it. After many goodbyes, we set off again.

This time we had to find our own path through low-growing palms dwarfed by massive vine-entangled trees. After some hours of slow progress, another moment of truth arrived. A tree had fallen across a deep ravine, forming a bridge with no handrails and a smooth, slippery surface. The first Dayak skipped across and then the second, and then I gingerly made my way over. On the far side Peter handed me two cigarettes. 'Start chewing,' he said, 'because we're in leech country.' Within an hour there were half a dozen leeches on my body. To get rid of them you had to drown them in cigarette spittle till they fell off. They seemed to come from the low, wet, muddy grass we were wading and pushing our way through. At least the excitement of being eaten alive made us forget how tired we were.

Back on the riverside, clear silvery water rushed over dark-grey stones that looked like polished cobbles. The crew dashed across, one with the outboard on his shoulder. It was comical to watch them as they danced and cavorted their way through the shallow torrent. Then it was my turn. To the amusement of my fellow travellers I went head-over-heels after five steps. I stood up, dripping, and made a run for it. It was easier upstream than down, but it still took me three times as long as the others to make the crossing. This was only the beginning. Then came a high, steep mudbank which the crew walked up, feet and toes spread like ducks'. I tied my trainers round my neck and went barefoot, medical advice long forgotten. It was up the bank or stay the night. Up, of course, was followed by down, then up again. It was as if somebody had ploughed giant furrows for us to cross. Even Peter was showing signs of wear. 'Not far now,' he would mutter as his men moved on, unperturbed by their light jungle ramble.

At the next longhouse, the inmates again streamed out to meet us. After the formalities, another development officer who had also trained in Australia talked to us about the schemes he had under way. There were small, flat rice fields, vegetable gardens and an orchard. His pride and joy was a large fish pond filled with water diverted from the river. After our tour of the premises we returned for food, more talk and games. Blind man's buff was a real winner. Collapsing at last onto a hard mat at the end of a full day was most welcome.

At dawn, despite the downpour, our enthusiastic officer was up and insisting on a swim before breakfast, so we dived into his ice-cold pool. It was a bracing start to the day and, inspired by the energy of my new friend, I was ready for the next long tramp. This time the way was steeper, the rivers faster and deeper and the trees taller but less choked with ivy and liana. We were close to the border now; this was the heart of Borneo.

By mid-afternoon and after many difficult crossings we saw our next longhouse. This time there were no pigs, only a few hens running under the house, and a few tethered cocks. A wizened chief conducted the formalities and discussions began.

After inspecting the rice stocks, Peter calculated what the shortfall could be, so what reserves were needed as contingency. We walked to rice-growing areas: river slopes of pure mud. There was a distinct possibility of a washout. The fields were already some distance from the house. As we tramped back, Peter talked about diet and protein, stressing the link between food and health. He interpreted his dire messages so that I could follow.

Back at the house Peter set out their situation, at each stage of the conversation gaining a reluctant buy-in. He kept pushing the responsibility back on to the chief, asking him for solutions. Eventually it was clear that the chief, old and proud as he was, had to ask, 'What can you do for us?'

This was Peter's opportunity and he took it very slowly. 'You could lead the people to another, better place and we could help you.' There was strong reaction to this so we started all over again. Eventually the idea that two of the young men could go and take a look slowly took hold, and this was finally agreed.

The chief then asked me to look at one of his sons who had a huge gumboil. Rummaging in my trusty medical kit I found a needle, and having sterilised it with a flame I burst the abscess to release the pus. I pressed until clear blood flowed, which I staunched with gauze smeared with antiseptic cream. The boy's relief was obvious. Now I was not only a managing director but also chief medical officer.

These roles carried much responsibility, I discovered, when I was then handed a sick baby. The infant's stomach was as tight as a drum. I spooned castor oil into a protesting mouth and began to massage the stomach, hoping to manipulate it back to normal.

In that semi-dark room sitting with a small baby on my lap I began to muse. What was I doing in the middle of Borneo among head-hunters with a life in my hand (not even hands), basing my medical treatment on a First Aid Proficiency certificate from the Boys' Brigade? A teaspoonful of warm water and three hours' gentle rubbing later, my patient emitted the gentlest, quietest fart—the first of many as his stomach relaxed. Crisis over, I rejoined my colleagues to feast on the staple food of rice and ferns.

The medical dramas behind us, we started to unwind, finishing the evening again with gales of laughter, crashing into each other in the most uninhibited game of blind man's buff. Women as well as men were involved this time. In the morning the baby's parents came to give me a precious egg for my breakfast. With solemn thanks I said it should be kept for the baby—but served soft- not hard-boiled! As we left, it occurred to me that they had never seen a white man before and perhaps they would assume whiteness was connected to one-handedness.

The crew met us with the prahu at the head of the river to take us to Lubok Antu. The river was fast and the rapids exciting. The bowman skilfully fended off rocks while his crewmate in the stern ran the engine. At one point as we negotiated a mini-ravine at great speed we heard a crash, and just ahead of us a wild boar leapt out of the bush into the river. With a swoop of his parang the bowman all but severed the boar's head, then he grabbed its feet and hung on as we swept down the river. To add to the drama a shot rang out, a bullet whistled over our heads, and an irate hunter emerged, yelling at us from the bank. A rapid exchange in Dayak extracted a promise that we would deliver the boar to his house downstream.

That evening we sat down to burnt pig and, believe it or not, ice-cold beer. We learned that the owner of the boar and the attractive house we were sitting in was also the holder of the Military Cross, gained in the Korean War. His citation had mistakenly located him in Indonesia, which would have been unthinkable, so to rectify the situation the government had given him his own house in Sarawak. It was a memorable evening with many heroic re-enactments of the hog-in-the-river drama. But despite our urging, our host would say little about the action that earned him his Military Cross.

After rice and ferns with a carbonized pork chaser, my stomach was ready for something more palatable, so having dropped Peter at Lubok Antu we stopped at Engkilili on the way back. Food that had seemed unappetizing a week ago (a plate of vegetables and a greasy egg, which I had politely declined) was now not only palatable but positively mouth-watering. Hunger conquers all, and from that moment onward I was great deal more flexible on cuisine.

*   *   *   *   *

Our next trip was with the agricultural officer. We went to the coastal areas where Dayaks were wealthy as a result of rubber plantations. They were self-sufficient, having rice fields and gardens with pigs and chickens in abundance. We talked not about survival, but about agricultural investment and politics. The walking was easy but the drinking was hard: these men knew how to party. The story goes that when their leader went to Singapore he refused to go to bed as long as the lights were still on in the city. Despite lengthy explanations, he would not be persuaded to turn in for the night, obliging his hosts to go out again and again to find the party. No embassy head ever had a more exhausting guest.

At Easter, Don and I took a break and went sailing on the Batang Lupar. There were tropical beaches where we fished and cooked the catch. We drifted up channels to small villages, picked up Chinese meals and then went out on the tide … and we were being paid for this!

Before we left this idyllic place, we had a farewell dinner with Griffin. This time he asked the questions, and as we answered he filled in the background. He knew the division like the back of his hand. The time we spent there was enormously worthwhile: building a solid base of experience, understanding a new culture and witnessing the process of change; seeing how to wait and when to talk; realising that common understanding is the biggest challenge, and having the patience not to move forward until that understanding has been established.

We spent our last night in Assistant District Officer Hal Ferguson's house. A serious man in his early 30s, he was in love with the country and with a beautiful Dayak woman. He feared that there may be communist intrusion into this undefended country, a concern that had also been evident on our coastal trip. Having seen something of the terrain and heard of the Dayaks' competence in defending their territory, I doubted the likelihood of a land-based attack, feeling that civil unrest was probably more likely.

To celebrate our last night, we had strung Chinese crackers underneath the elevated timber house, and at midnight we set them off. Unfortunately, we had neglected to tell Hal what we were up to, and the sound of them suddenly firing sent him into a panic. He leapt up, grabbed his rifle and pointed it into the darkness. We were terrified that he would pick off the night watchman, and then the gardener! The rest of the town slumbered on, unconcerned, assuming perhaps that the fireworks were some restaurant celebration. To our relief, no lights came on in the residence on the hill; Griffin had seen the real thing and he did not stir. We carefully disarmed Hal, gave him a brandy and apologised. Next time we would be sure to mention any planned pyrotechnics well in advance!

*   *   *   *   *

The return to Seria and the oil camp brought me back to comfortable suburbia and a new assignment. Taking over from Bill Bentley, an intense, flamboyant man, was a challenge, but working for Barney again was great.

When news came through that the Duke of Edinburgh was about to visit, someone suggested that Barney should be given free rein to design the reception. We thought we'd better have a dry run to make sure all went faultlessly on the day. The practice entourage had hardly left the airport before Barney turned left into the mangrove swamp for a quick look at the crocodiles' entry point, and then left at the intersection down to the rubbish heaps behind the machine shops. On to the Indian quarters, where he conducted a walk through the living rooms, disturbing a sleeping man and a dog, and into the kitchens, which were piled high with dirty dishes. Then back to the car, into the marketplace, and on to the staff club for comparison. The route was subsequently wisely adjusted to turn right into the gas plant, not left into the dump. The Indian mess, just in case, was cleaned and painted (they were the winners in this charade). In the end the duke went straight to the office and then the club. He never knew what he missed.

Working with Barney was an education. Like Peter Tingom, he was good at listening. Always suspicious, he saw crime everywhere and was tenacious in his detective work. When the health officer complained about the domestic staff's chickens running around the *amah*s' (nurse maids') compound, strict rules were put in place to rectify the situation. But these did not work, and the amahs went on strike. The solution was to procure chicken coops, so a contract was drawn up—and the coops miraculously appeared in next to no time. The staff were happy and peace was restored ... but Barney saw a mastermind at work. 'It can't be the health officer,' he said. 'He's too thick.' So he devilled away, pursuing the origins of the contractor, distracted from time to time to deal with stabbings—disappointingly all superficial—and petty theft. But before he was satisfied, his powers of detection were required elsewhere.

The company cashier was found collapsed on the floor of his burgled office, all the cash gone. Security men ran everywhere looking for clues, but Barney simply walked to the club and asked to see the cashier's bar bill. There it was: scores of bottles of whisky signed for and then no doubt sold in the market for cash to pay his gambling debts. A stupid, desperate scheme. Ignoring protestations from the cashier's line manager, Barney had a confession wrung out of the man while the security men were still out looking for robbers in the backwoods. Jack Frost the TV hero would have been proud of him. Barney was never smug, but he got close that day … until I reminded him about the chicken coops.

He quickly changed the subject and told me he wanted me to conduct the post-vacation interview with a Malay unionist, Sulaimon, a notoriously difficult man from the water transport division. 'Your Malay will get you by,' he told me, and for good measure he asked senior official Pengiran Bahar to sit in.

Sulaimon arrived and presented his tickets with his application for reimbursement. The tickets were for Simanggang, and I noticed they were fresh and clean. Nobody comes off the Batang Lupar boat to Simanggang with anything fresh or clean. Even the meticulous Don had not been able to achieve that. Sulaimon and I exchanged pleasantries about his journey and his holiday, and as he began to relax I asked him what he thought of the new mosque in Simanggang. He agreed it was superb. We chatted for little longer then off he went.

Barney had waited to see how the visit had gone. I closed the door and told him it was a mega-racket. 'These are false tickets and this man has never been to Simanggang. There may be a basketball court there but there is no mosque.' Pengiran confirmed the exchange and Barney went into orbit. Chicken coops were forgotten as he jabbed the numbers on his phone. Within days the ticket-printing press was discovered, the owners and their agents arrested and our union man charged. The racket was closed down, but my man went free because the prosecution, in presenting its case, had omitted to identify him. David Marshall all over again!

*   *   *   *   *

Barney's blood pressure came down eventually, but rose again after Hector saw the headlines in the Hong Kong press with a quotation from my predecessor, Bill Bently. Bill was promoting close cooperation with Brunei and promising huge employment opportunities for the Hong Kong

Chinese. Hector—always sensitive—immediately saw the potential impact on Brunei, and hit the buzzer for Barney. Barney said, 'Sorry—Bill's gone. Can't reach him. He's exploring in South America, I think. I wouldn't waste time on it—he'll probably leave anyway.' He did, but not till 30 years later! Barney looked after his troops; he had learned the importance of that in earlier, tougher times.

These post-vacation interviews helped to fill out a picture of the employee landscape, as did the resignation interviews. One of the latter particularly stands out in my mind. A tall, straight-backed Chinese worker came in to talk about his decision to resign. He had worked for over 15 years as a cleaner in the clinic and his record was without blemish. We had recently amalgamated the three lower grades into one, at which point his pay had risen, since he had been at the lowest grade.

I listened carefully as my interpreter, Chye Moi Lin, translated the reasons behind the worker's decision to leave. The man had originally travelled from mainland China through Hong Kong to Brunei, and now he wished to return so that he could die in his homeland. Chye had tears in his eyes. As I signed off the papers I wished him well, hoping his return trip would be less arduous than his outgoing one.

Two months later as I drove down the main road I overtook a statuesque figure on a bicycle, pedalling along with his groceries in a basket on the handlebars. This was my emigrant—obviously not on the first stage of his journey to the homeland! The next day I told Chye I had seen our interviewee on his bike. With some embarrassment the story came out. Resignation had not been about a return to the homeland, but about the fact that the man had been grossly insulted by the amalgamation of the job grades and the quasi-promotion. He had not worked at and been paid at one level for all that time only to be told he was at the wrong level. His job had not changed, and yet he had been promoted. To him this meant that he had been being cheated for 15 years. He could not work with people who would do this. Rather than lose face by accepting this deception, he was implacable in his resignation.

My discussion with Chye moved on to the philosophical. Europeans, it seemed, too easily accepted the readily comprehensible, and in so doing missed, so did not address, the reality. Being persistent in pursuit of the truth was the key. This made me think and prompted me to work harder at getting to know my interpreters and colleagues from the East. But persuading them to trust you with the facts in a situation was not easy, especially because it often made the job of rectification that much more difficult.

Some decisions could be hard in this first real job. A case came up that required a decision, and unhelpfully there was no precedent. An Indian diagnosed with terminal cancer had been hospitalised. The end was not far away. In hope, he had requested to return to India. He had already tried booking his own passage, and failed. I bypassed the usual channels and organized for him to put on a plane the next day. The tickets were issued, but not before the excellent chief clerk, John Matthews, pointed out that such a decision was not technically within my remit. I assured him that it was my neck on the block. I could see the case being referred higher and higher through the administrative ranks until the terminally ill man had reached his final destination without fulfilling his last wish. This, I said, was not the way a multimillion-dollar company should proceed. The next day, the patient left.

Twenty-four hours later I was horrified to hear that the plane had turned back on its way to India and our man was hospitalised, now in Singapore. Thankfully, at almost the last gasp he had re-boarded the plane and arrived in India in time to be with his family. The messages from Singapore did not make good reading, but the letters overflowing with praise from our representative in India did. They were addressed to a slightly bemused MD, who forwarded them with a 'well done' note to Barney. Barney in turn passed it to me, asking 'Was it?' And that's where we left it, but the principle was fundamental to me—I would go on as I started. I would make decisions!

This serious episode passed into the next trivial dispute. Two amahs had indulged in a bout of fisticuffs over a rather unattractive cook. The interpreters began by reporting the name-calling rather than the violence, because that was apparently where it had started. Both women had dressed smartly for the interview and neither showed any signs of irreparable damage. The first inflammatory jibe had been thrown by the senior nurse who, the redoubtable Chye translated, had called her rival a 'slim chicken'. Here was where language misled. By careful questioning 'slim' changed to 'scrawny' and 'chicken' to 'hen'. This oral interplay between the interviewer and the interpreter did nothing to ease the tension, so instead we sought common ground. They both liked their jobs. They were well paid, they loved the children they looked after and they did not want to return to Hong Kong, so the threat of punishment dissipated. After this confrontation with reality, everything calmed down and there were no further wrestling matches or visits to the office. Indeed, whenever I subsequently passed them shepherding their charges, they would give me a smile and a wave.

Life in the labour office was never dull, and as an introduction to multiracial living and working it could not have been better. The workforce was drawn from all parts of the Far East, and many—as youngsters— had suffered the terrors of war and occupation. Because I had travelled and lived with Dyaks, some colleagues assumed I was an authority on communicating with them. One scorching afternoon I was called by an exasperated Australian to come down and talk to 'these friends of yours'. It was about a painting job that was 50 or 60 feet up. The men were merrily working away in flip-flops rather than safety shoes, and without safety harnesses. Progress was excellent and painting quality first class, but rule-book recognition was zero. I described to the Aussie how I had seen barefoot Dyaks trip effortlessly across trees bridging ravines far deeper than 50 feet. They simply did not think about falling, and we could not satisfactorily answer their simple question, 'Why would we fall?'

After a discussion with the senior painter about carrying a pig over a ravine (which diversion really excited the good Aussie), we sort of got the message home. They accepted the safety line—but not the shoes. They explained that they needed to feel what's under their feet, and flimsy flip-flops allowed that.

The story spread fast, as did another proving that the Dyaks had their own way of doing things after a Dutch engineer employed one to dig his garden. The engineer showed him the flower-bed, gave him the spade and left to play golf. Two-and-a-half hours later he returned to find the man 15 feet down, dangerously close to the gas structure that ran close to the surface in Brunei. The digger was as mystified as the engineer was amazed. He had said dig the garden. A hole is a hole: that's what he wanted and that's what he got!

Another incident that cemented respect for the Dyaks involved packs of vicious wild dogs which periodically ravaged the camp. Our attempts to shoot them had been both dangerous and futile: one bang and they disappeared back into the Bush. The Dyaks' accuracy with the silent blowpipe was legendary. After two Sunday afternoons' work, the dog problem had disappeared.

These lessons about communication and cultural divides were invaluable. We in our multiracial cities and societies must learn that if you take the time and persevere to understand your neighbours and their talents, the dividends can be huge.

*   *   *   *   *

The platform that the men had been painting was destined for Ampa Patches, a good half an hour by helicopter from the mainland. The enormous structure, effectively an offshore island, would be towed there by tugs and then positioned over the site. Construction of the drilling rig began immediately, and after it was finished, the blowout preventers were positioned over the hole. It was an impressive engineering achievement. But as it was being constructed, it became clear that the organizational structure behind it, which crucially had been devised to maximize efficiency for projects on dry land, did not work. There was a surplus of non-relevant expertise. As the project developed, the numbers attending planning meetings grew until the conference room could no longer hold them all and the meeting had to be moved to the canteen!

The offshore operation needed a workforce with specialist skills tailored to its particular challenges. We also needed to recognize that the two phases of the operation—the construction and the installation—were separate, and called for different skills. We had to define our exact needs not only for effective execution, but also for cost control. Watching a project evolve from being based on land to being operated out at sea was not only fascinating, but also an invaluable education and preparation for my later responsibilities.

The physical difficulty with Ampa Patches was rooted in its geological structure. When the drilling got under way and the target was within reach, the drillers encountered their worst nightmare: heaving shales. These are like melted toffees. The drill bit cannot make headway and effectively changes from being an incisive tool into a mixing machine, circulating the gloop and going nowhere.

When our great out-step venture failed, the camp was despondent. The fact that we had learned how to organize and build it was little consolation. With hindsight we could see that a stationary platform was too big a bet. We needed to be able to change locations without rebuilding, which meant mobile drilling rigs and then fixed production platforms. What was also becoming clear was that the drilling platform was an attractive area for the contracting sector. There was no reason why we should invest in the drilling activity when there was a contracting industry ready and keen to be involved.

For me it had been another learning experience, not only from the angle of organizational fit and change, but also from the structure of investments necessary to finance what was in fact a new industry. It was like moving from cars and lorries to helicopters and aeroplanes!

We single men started work early and played late, with lots of physical activity in between. The more organized houses had dinner parties and gatherings. At our end of the bungalows we had invitations but empty fridges because we had forgotten to shop or to give the cook the funds for the market. But we did not starve, and we even learned how to shave on the way to the office—which, with just one hand, is quite an accomplishment.

The camp was full of characters. Among them was George Brown, a senior member of the drilling department. George was a big, massively strong person; his party piece was to reach up with one enormous hand and stop the overhead fan. His wife was a minute Australian who took no nonsense. When she went off for a three-month trip, a strange thing happened to George. He took in a lodger.

He had found a deserted bear cub in the jungle near his rigs and, since he now had the house to himself and had the space, he took it home. The bear loved George and George reciprocated. They rolled around the house in continuous wrestling matches and George's ripped shirts and clawed forearms testified to Baby Bear's growing strength. But when Mama Brown returned, Baby Bear was given its marching orders and dispatched to a California Zoo.

A few months later, George diverted his homeward passage to California and an old friendship was renewed. To the curator's amazement and alarm, George and No-Longer-Baby Bear hugged and cavorted around the cage. On his return George was happy to report that the bear knew him immediately … but it seems it did not know its own strength. George liked to think the encounter had ended in a draw, but Mama Brown described it as more of a rescue operation.

*   *   *   *   *

I had initially been told that it would be three years before I could bring Jo out to join me. However after a year, by which time her lectureship at St Andrews had ended, I had convinced them I was not a short-term prospect. I requested permission from Barney to go to Singapore to be married, and he granted it straight away. Getting married in Scotland would have been much easier, with all the support systems on both sides of the family, but it was not to be since neither of us could wait. I did not want an oilfield wedding where I knew everyone and Jo knew nobody, so we decided on Singapore. When my friend Ken Meyer, an optician based on the island, heard of my forthcoming nuptials he offered to—or rather insisted he would—organize the reception.

I chose a church and Jo had the bright idea of advertising for guests, describing our parents and where we had come from. Ken loved the idea, and it worked out amazingly well—the venue was crammed with well-wishers. Ken's gift was a wedding cake which, when you opened it up it, was filled with beautifully packaged slices, all ready to be addressed and sent to friends and family at home. It was a typical Meyer touch. Jo, the most beautiful bride, was delighted with the arrangements—even with the funeral lilies I had ordered to adorn the church! By the time I had finished my game of golf that afternoon and remembered that I was supposed to have arranged the flowers, the lilies were all they had left in the shop. The festivities went on long into the night, and as Jo said it was real fun because we had done our own thing. The company plane picked up two weary but very happy newlyweds early the next morning. We moved straight into our kajang house—a traditional wooden house with no windows and no doors—by the sea. Jo took it in her stride and quickly turned it into a home.

* * * * *

I took up a new job as assistant to the technical manager, which was closer to the business. An oilfield is resource-heavy when it is establishing itself, but when it moves into the production phase, resources need to be streamlined to make sure they're cost-effective. It's a major operation.

I started by asking for the manpower forecasts from each department. Their assessments made it clear that there was no appetite for reducing numbers, no willingness to challenge or change. This was disappointing, so we projected a much lower set of numbers and asked what activities would be curtailed and what would not get done if the teams were reduced to these sizes. This was not popular, particularly when we followed up by asking why certain procedures were necessary at all, and whether there were more efficient ways of achieving the objective. It was slow going.

The lesson I learned as I coaxed and struggled with the Australian workshop people, the Dutch drilling superintendent and a fairly arrogant engineering team was that the turning point comes when the initiative to change comes from the operators themselves. Beneath the grumblings and the protests there was a hard realisation that change was linked to survival. It would have been easy to back off and point upwards saying,

'Don't blame me, it's what the boss wants', but I was not prepared to do it that way because I believed we were massively overstaffed and extending our role far beyond our remit.

Barney had warned me that I would be meeting some strange people in this new role. He had a list of them, complete with colourful descriptions. There was one particular man in production who, when he arrived, called everybody in the department together and declared, 'From today, production is under new management. Things have got to change.' He then listed the manpower reductions he was intending to make. This would have been good news had he been in charge of a department that was key to the development phase, where staff were no longer needed once the job was done, but in the production phase, your trained team was an increasingly valuable asset to the operation.

The new commissar rang me the next day and said that one of our training programmes was misconceived and he was cancelling it. Twenty-five trainees would go. I talked to Barney the Wise about it and he counselled me to wait. 'There will be an announcement of a new programme to make good the short-sightedness of this,' he said, and before long there was. The 25 trainees from the cancelled production programme were now official corporate trainees attached to the personnel department. We had not lost our trainees—although the new production manager had managed to shift them onto someone else's payroll!

*   *   *   *   *

It was Melbourne Cup time. In Brunei, this was a 72-hole golfing competition held over two weekends, the second timed to coincide with Melbourne Cup weekend in Australia. In Melbourne, the Cup was (and is) one of the world's great horse races on which almost every Australian family is likely to have a bet. It was the only topic of conversation amongst the Aussies for weeks beforehand, it generated feverish excitement and you could not help but be drawn in.

The Australian High Commissioner visited our Panaga (sports and social) Club and presented us with a replica of the original cup. His speech paid tribute to the Australian soldiers who had played a vital part in the liberation of Brunei. The gift came with only one condition, laid down by the Melbourne racing club, and that was that the replica was never to re-enter Australia.

In our tournament, instead of backing a horse you bid for and 'bought' your player. It was to be a handicap competition so every competitor had a chance, and the players' form was already under careful scrutiny. Barney was in the lead in assessment. Two of his competitors had been posted elsewhere so the field was open. Tension grew, as did the stakes.

A quiet, almost silent, man named John, who was a powerful driller, hit the ball a long way, but not always in a straight line. Barney had christened him 'Johnny the Sphinx' and his recent form was good. The next horse was Phil, who was not quite at home on the green and was certainly no racehorse. He was a very competent sailor, though, so Barney wondered if the plentiful water on the course might boost his confidence. Then there was Mike, a fastidious bachelor who hit a long ball and chain-smoked throughout the 18 holes. Barney rejected him outright, saying he was too keyed up to win a major. I suddenly realised we were in the big time!

The dark horse, geophysicist Geoff, was the thinnest man I have ever seen. He worked all day and slept all night in an air-conditioned room, but Barney didn't think that that compromised his chances of taking the heat—physical or mental—of the competition. The bets were on. I went for a few dollars, mostly put up by the schoolteachers. Every day Geoff found water bottles on his desk, along with advice on preventing heatstroke. An anonymous donor provided Phil with a map locating all the streams on the course, with instructions in Malay to stay away from them, since it was crocodile mating season. Johnny the Sphinx had regular visits from his owners. Popularity was a new experience for him and he began to enjoy it. He even started serving coffee.

On the last day's play, the heat began to tell. The Sphinx had a bad one. Geoff had drunk so much water that he was afloat and not connecting. Barney was well placed but under suspicion that he had deliberately spooked the opposition. Phil had ignored the map and the warnings and almost run out of balls. The 50 cents originally invested in Lee the Lean Australian had crept up, and it was he who came into the final straight in good shape and won. The Melbourne Cup, Borneo-style, had provided just what an oil camp needed: entertainment, exercise and ridiculous fun!

Lee the Lean might not have won his heroic victory had the two most promising candidates not been called away at the critical moment. Bert Blanche had been on a break in Hong Kong. When he returned he was keen to show off his various acquisitions, among which was an electric carving knife—the height of chic gadgetry at the time. It worked a bit like hedge-trimmer and Bert proudly demonstrated how one could use it to amaze one's guests by slicing, this way and then that, through a block of multi-coloured ice cream. His housemate, Arthur Buchan, had missed the tournament because he had had to go back to the Bush to drill another hole. Later, over variegated ice-cream slivers, he told us how his mother had been writing to ask if he had received his Christmas present. He had visited the post office several times in vain, so while he was away he had asked his cook to pursue the matter. On his return, the cook presented him with the present in a bucket. Arthur opened the dripping parcel to reveal a beautifully knitted pair of woollen gloves marinated in a pound of his favourite, now rancid, butter.

*　　*　　*　　*　　*

I had had the job of golf club secretary foisted upon me. The club was easy to run, that was not the problem; the clubhouse was another matter. It was old and deeply loved, particularly by the Australians and more particularly by Barney. The company had built a new recreation club at the opposite end of the course from the clubhouse. The idea was that we would migrate into these facilities and reverse the numbering of the holes. 'You can't start with a short hole,' diehards objected.

'But didn't Royal Lytham, a regular host of The Open, not have a short first hole?'

'Whose side are you on?' was the grumpy response. This was rapidly becoming a cause célèbre. It was like student union politics. I picked up the gauntlet and decided to use the unpopularity of the move, which in that closed community had begun to worry senior management, to secure the best possible facilities in the new club: changing room, bar and club room. It worked. We were granted all we had asked for, and the move gained the full support of the committee, including Barney, who was now captain. The general meeting amid many libations approved the move, and to everyone's delight George performed his party piece, reaching up with one giant hand to stop the overhead fan! We could all get back to the serious business of golf.

*　　*　　*　　*　　*

Back at the plant, now that we had rationalised the manpower, the time had come to take stock. It was obvious there were things we did not need to do for ourselves. There was sufficient expertise and capital in the local economy to take care of our vehicle maintenance, for example: to supply the vehicles, even, on lease or on contract. The process of change started, and slowly we began to withdraw from being the universal provider and return to the primary activities of finding and producing oil.

A seminal group paper arrived from London for me to review. The author, A. P. Blair, asserted that the Group would maintain its momentum only if it recognized the need to involve local populations both financially and actively. This would not only give us access to new talent but also build a strong body of support in the countries where we operated. This policy fitted well with the independence movements already under way in the old colonial empires.

The formula for optimum staff numbers was 100 - x, where 100 represented local people and x was the cadre of expatriates who would initially facilitate the transformation. The x component would gradually reduce to a smaller proportion in order to bring the advantages of the international group to the domestic operations. Eventually these expatriates would come from the ranks of the national companies as well as from the main centres in the UK and Holland. It seemed so obvious, yet in 1958 it was a revolutionary plan. It was a mantra I passionately believed in and dedicated myself to applying for the next 30 years. The Group has now realised this target, which has been a source of great strength.

In Brunei the situation was complicated by the range of people involved, drawn not only from Europe and throughout Asia, but also from Borneo itself. I prepared a paper setting out the nationalities we employed, and sought advice from our host government as to its acceptability. The government and the sultan, well served by their advisers, responded clearly. This was a country for the Brunei Malays and everybody else was a visitor whose welcome would be determined by their residential intention. The closer their country of origin, the less acceptable they were, seemed to be the line.

My comments on Blair's paper were, first, the importance of listening to the government's advice, and secondly that we needed scholarship schemes aimed at engineers and scientists in each of our territories. We should also be trawling the best universities in the UK for potential recruits, as English was the language of education in Borneo. This was agreed, and I was instructed make a start. The schemes were welcomed in Sarawak and in Sabah, but although the schools were helpful, the exceptional candidates we wanted were not easy to find.

I had identified one brilliant Chinese boy called Yee Chee Pong. He had left his school in Kuching and, after some detective work, I discovered he had come back to Brunei where his mother was a washerwoman. I asked around and one day he just turned up in the office, bright-eyed and spiky-haired, saying, 'I hear you've been looking for me!' When I described the scholarship scheme to him he said he would like to do it, but first he wanted to join the artisan scheme in the workshop. That can be arranged, I said, but warned him it was a two-year course. 'I will finish it in three months,' he replied. He started the next day, and finished with distinction within the quarter.

Yee Chee Pong's determination, I learned, was partly inspired by the industrial accident his father had suffered in the workshop, the effects of which had cost him his job. He wanted to know where the dangers lay in the work and why the accident had happened. He also wanted his father to know that he had learned his trade.

Another of my students was a Malay, Mansor bin Kipli, a quieter person than Yee Chee Pong but a brilliant electrical engineering student. He became a senior engineer and was selected to accompany an offshore platform on tow from Hong Kong. Caught in a typhoon, the platform capsized and all hands were lost, among them Mansor and his brilliant future. Catastrophe in our business was the permanent backdrop of our lives and no respecter of talent or prospects.

Jo had found a teaching job in the nearby town of Sema, which was enormously rewarding. Her pupils, mostly Chinese and some Malay, were eager, bright, and fun to teach. She introduced all kinds of new reading material into their curriculum and promoted historical studies.

Somewhat to our surprise but to our great delight our first son, James Douglas, was born in 1959: 'Ah Jim' as the Chinese amah called him, although he was always Douglas to us. Jo would feed him in the middle of the night in his bedroom, which overlooked our sociable French neighbours' house. Very late one evening she was amused to see their houseguest, a large helicopter pilot, trimming his beard and pointing up his moustache. He called a cheery *au revoir* as he set out for his nightshift on the platform just offshore.

We were woken at dawn by helicopters and planes flying low over the sea. It was not difficult to guess that something had happened; it turned out that the helicopter had crashed due to engine failure, and the elegant Frenchman was no more—another reminder that this was a dangerous industry.

By the end of 1959 Yee Chee Pong had left for the UK and I had been reassigned to Nigeria and was in London. One morning I received an anxious call from the central training department. Yee Chee Pong had not come off the plane. Learning that it had been delayed for 24 hours in Bombay, I asked the woman if she had tried his university residence. The woman thought this was a facetious question but I said it might be worth a look. She rang me back very relieved. Impatient with the delay, Yee had changed planes, flown to Frankfurt and then on to Heathrow to arrive in good time. He finished his studies—again with distinction—and returned to Brunei. He later emigrated to take up a professorship at the University of Western Australia.

*   *   *   *   *

In Borneo I had by good fortune had a fantastic start in the oil business. It was not a narrow, textbook start. It was set in a changing political environment and I had felt the growing tensions that culminated, only months after our departure, with the Brunei revolt. My introduction to oilfield work had included seeing and learning how a colonial government works, and living with people who survived in remote places in the Sarawak jungle.

Just as we were about to leave, the three-year-old son of our close friends drowned by falling into an uncovered well head filled with oily water. Awareness of risk is all about personal discipline and all of us who walked on that beach had been negligent. I can still see my friends standing in mourning dress in the tropical heat, and the priest trying to share the parents' grief. The child's mother had been at school with Jo, and as Jo clutched Douglas that evening I could see and feel what she felt, and the question she might be asking herself.

Our three years in Brunei had been an important formative part of our lives. We had begun a loving partnership that was to endure into our 80s, and with the birth of our first child we had become a family. We had survived camp life, a feature of the oil industry for well over a century, which can be claustrophobic or lonely and miserable, especially for women whose husbands are absent for long periods in the Bush or offshore. Survival skills sounds like extravagant terminology for what was required, but you could certainly see when they were lacking. We had come through our first test in the life we had set for ourselves, and were braced for the next.

3

Nigeria

During my three years in Brunei I had seen colonial administration at work, with a rudimentary justice system and a reasonably effective security force. The move to independence in Sarawak was already under way, and different groups with diverse religious affiliations and political aspirations were included in the process. In Africa, the winds of change were blowing and we were about to experience a unique period of political history. Jo, who had studied in Virginia in the US and had sat in the 'wrong' seat on the bus as a statement of her convictions, had more immediate insight than I did. We set about studying the country, its peoples and the history of their political development. With independence approaching, the future looked exciting.

The size of this country and the growth of its population were staggering. The Niger set a boundary between North and South and East and West, areas separated not only in geography but also in language, personality and political structure. The colonialists' arrival had brought a common language, English, which facilitated the development of a judicial system and the establishment of international trading houses. The Christian church, an important force in religion, was arguably even more important in education. At primary and secondary levels, Christian-influenced education was concentrated in the South. Its spread to the North was limited in part by geography but more fundamentally by the Muslim faith.

The physical creation of Nigeria in the nineteenth century was the work of Lord Lugard, who had served his colonial apprenticeship by securing treaties with kings and chiefs while warding off French intrusions. Lugard created the protectorate of Northern Nigeria by conducting successful military campaigns against the Emir of Kano and the Sardaunas of Sokoto. A few years later he combined the protectorate of the South with the North to create modern Nigeria. His previous experience in East Africa shaped his policy of indirect rule and land rights for the indigenous population. Indirect rule left the Muslim North in the hands of the established hierarchies, the sardaunas and emirs. These feudal structures would have the biggest transitions to make to meet the demands of modern democracy, but they were reinforced by their commitment to the Muslim faith. The North was more cohesive than the freethinking, politically chaotic South.

In the South there were no hierarchical power structures. There were chieftaincies, councils and village elders—a hotchpotch of structures sitting above highly educated lawyers, teachers, businessmen and farmers. Journalists ran papers that espoused a wide range of politics and philosophies, the editors fighting libel cases from inside as well as outside jail.

The land from Lagos in the West to Calabar in the East was studded with churches of many denominations. Catholic priests tended their flocks in some cases with an intensity which, within a few years of independence, would explosively and radically influence the country's future.

The management of Nigeria under colonial rule worked because there was a referee who had no political interest to protect or advance. After independence, when the referee disappeared and the players took over, the mechanism to make and execute trade-offs disappeared, and with it the middle way, through which the country had slowly and peacefully been making progress.

*   *   *   *   *

The arrivals hall at Lagos Airport was a teeming mass of people and a jangle of vibrant colour and noise. Huge cases, new tyres, sewn sacks were hefted away to waiting cars. Laughter abounded amid spikes of argument. Reunited friends and families embraced, nimble porters seized luggage only to be shooed away; babies slept soundly on their mothers' backs. A Dutch painter could have made a masterpiece of this mayhem. I was travelling alone; Jo was following by boat, bringing baby Douglas and the car.

I was to fly the next leg in a company plane with the managing director. This could have been part of a cost-control policy, but more likely it was the MD taking the opportunity to check out who was coming to join his family. He sat upfront and shouted over the engine noise: 'We'll fly low so you can see what you've come to!' I looked out over the Niger Delta, a spaghetti-ball of rivers making their confused way to the sea. He had opened with, 'So you're the golfer—welcome!' which I brooded over later. When I won a scholarship, the whispers were that it was the golf. At St Andrews it was golf and union affairs—but it was also work. These associations are among the currents that direct life, but it is disturbing if they seem to overwhelm other achievements and aspirations.

The wet season was in full spate as we descended into Port Harcourt. I could see nothing but the outline of palm trees as the rain streamed down the windows. We taxied to a stop outside a large bungalow that served as the airport building, and ran from the plane to the door. It was like walking fully clothed into a warm shower.

At the compound the welcome was friendly, as oil company welcomes always are. I made my way to the guesthouse that was to be my home for the first two months. The village, Umuokuroshe, lay outside the major town of Port Harcourt, which was named after a famous colonial minister.

A typical oil company encampment, the houses were well built, the roads and drainage ditches laid out in a logical way, the sports and recreational facilities centrally located, and there was a primary school nearby.

The African villages around this tidy town seemed to lack any logic at all. They had few identifiable roads, just a multitude of muddy pathways dotted with palm trees that were tapped for palm wine. Men and women rushed to and fro, oblivious to the pouring rain, skating on bare feet along the treacherous trails, washing babies in the dark-brown streams … there was colour, movement and noise everywhere. This bustle of people dressed in everything from rags to flowing robes gave a vivid impression of Dickens's London.

Port Harcourt, like Umuokuroshe, was built on dry land. The Niger Delta was on our front doorstep. The towns and villages lay on fingers of land among the swamps, which stretched away to the sea. Each had its own administrative hierarchy, from family leaders to chieftaincies, and more recently, in response to political progress, new village councils. Power was all about land and wealth. Land was your base, and wealth grew your base. This was the foundation of Nigerian politics. I saw it early at first hand through the eyes of a brilliant student who came to work with us during his vacation. Sonny Oti, aged only 21, wrote and performed in a play that was staged at our social club. The drama was set in an atmosphere of election fever when people were energetically supporting their candidates. The contest boiled down to two people: one a brilliant young orator bristling with idealistic plans, educated and smartly be-suited, and the other an overweight (heavily padded), loud-mouthed inarticulate dressed in traditional robes, sandals and a floppy hat. He seemed no match intellectually for his opponent—until the election began. Then he produced sheaves of notes from beneath his *agbada* and distributed them among the voters, who were clearly unfazed by his sartorial eccentricities. The outcome was inevitable.

This performance in the early '60s was acutely perceptive and prescient: a pointer to the reality that would emerge over the next ten years. The intelligent young were aware but unable to stir the conscience of a public that was out looking for *dash* (money). Dreams of independence were still alive, but already nightmares were on the horizon.

Independence took place in stages. The regions with their own institutions gained it first, followed by National Independence on 1 October 1960. Regional political activity had homogeneity about it— the tribal groups had ascendancy in their areas: Hausas in the North, Ibos in the East, Yorubas in the West and Binis in the Midwest Region. Consequently, political selection and decision-making was not complicated by multiple languages and dialects. National politics was an entirely different matter. The major tribal blocs set about making alliances, and so the seeds of trouble were sown. The process was new and immature, and the stakes were growing. This was no longer a groundnut, palm oil and cocoa economy; this was an oil economy and we, in the oil industry, were right in the middle of it. In a way we recognized what was happening more clearly than our hosts did. After all, we had seen it before, in Venezuela, Mexico, Iran, the Middle East and Indonesia.

Meanwhile, we had to get on and develop an oilfield. Production was already 30,000 barrels a day but exploration had moved to the swamps, where the key technology was 3-D seismic. Seismic helps identify and define the oil traps that lie under the surface; its effectiveness led to exponential growth in the Niger Delta. The workforce was a virtual United Nations, with Nigerian drillers, Dutch tool-pushers, British petroleum engineers, American welders, Swiss geologists, French well-testers and Australian field engineers.

Motoring in outboard-driven dinghies through the swamps and cutting the mangroves to set the seismic lines was a difficult exercise, but it paled into insignificance when compared with moving drilling barges—massive floating ships—into position so that drilling could begin. The terrain was hostile; malaria-bearing mosquitoes were ever-present. Strange viruses such as denghi fever attacked the body, bringing on throbbing headaches, high temperatures, rashes, and exhaustion. Weather conditions were also often foul, but success was immediate and production volumes shot up as soon as the pipelines were in place.

The swamp-drilling programme was determined by the rise and fall of the Niger, which at first we thought simply followed the wet and dry seasons. We realised there was more to it when waters fell unexpectedly, leaving our barges high and dry. Needing to be able to predict levels more accurately, we looked harder and further afield for the causes, eventually identifying them north of Nigeria in Mali, where the river itself rose.

As explorations succeeded and oil programmes multiplied, activity levels soared. Finding an oilfield is a fantastic experience. Success boosts everything: attitudes change, problems become solvable and everything forges ahead. But harvesting the earth's riches is not straightforward. She will not yield them easily and she will surprise you with her violent reactions.

It was Christmas morning when I awoke to the roar of what I thought were jet engines. The Congo had been in the throes of civil war these past few months and I thought I was hearing the sound of high-powered transport planes. But then the phone rang. There had been a blowout, and all hands were needed to control a perilously volatile situation.

This was bigger by far than my first encounter with an out-of-control well in Brunei. The roar of the fire, the heat and the trembling ground made people keep their distance. The question on everyone's mind was: How do we tame this monster? An oil company team has little experience in subduing blowouts because oil wells are not supposed to blow out. Rigs and wells are designed to prevent it from happening. Expertise is invariably flown in, and ours came in the shape of Mr Red Adair.

A red car awaited him at the airport … but he emerged a sick man. He had contracted a fever further north, probably in Libya, where for some weeks he had been trying to subdue an inferno. After a brief discussion on site he disappeared to hospital—and the fire went out. The local chiefs immediately agreed that this man was what was needed: he had exercised powerful juju from his red car and now he was resting until he was ready to clear up the mess of tangled steel and leaking gas. The site was now at its most dangerous.

Within 48 hours he went to work. He surveyed the array of engineers. He instructed the Australians to build two Olympic-size swimming pools at the edge of the site. He asked the mechanical engineers to rig up an apparatus that could pull the mangled steelwork away, and he called for cutting equipment. Explosimeters, which detect the location and measure the volume of gas, were put to work.

As the red-suited Adair made his way through the debris and scorched earth around the rig, cautiously probing the mangled remains, he passed a single palm tree. A small, muscular farmer wearing just a loincloth sprinted past the patrols and up the tree to fill his *calabash* (gourd) with palm wine. This was no landowner staking a claim for damages, it was a thirsty Nigerian wanting a drink. The ill-matched pair made a memorable picture.

Slowly but surely the steelwork was removed, and in two days the site was cleared and the well capped. At the end of it all, the Australians asked what the swimming pools had been for. Red said, 'In a crisis, the danger lies in men of action who act before they think and get in the way. You can't tell them that, so you have to give them something to do.' He had done just that, and although the Aussies were furious to discover their mission had in fact been just a ruse, the mood did not last. They had seen a consummate professional achieve a spectacular result.

The blaring fire engines, the turmoil in the Congo and stories of the high-flying U2 aeroplanes triggered thoughts of our eminent chief geologist, Erdi Frankl. He was still in Owerri, our first camp in Nigeria, which was slowly being closed down as we moved on to Port Harcourt. In addition to his technical responsibilities, Erdi was in charge of the camp. It had its own airstrip that took Twin Pioneers, the short-landing-and-take-off aircraft. He sent a confidential circular to his Dutch, German, Swiss and British colleagues advising them that one of the U2 aircraft, on its way to the Congo, would be refuelling at Owerri, and they might be interested to see it. It was due to arrive at noon. On the appointed day Land Rovers parked in the searing heat while their drivers peered into the sky looking for their first glimpse of this mystical Star-Wars aeroplane. Nothing happened until 12.30 p.m., when the geologist arrived in a jeep loaded with crates of cold beer to celebrate April Fools' Day. To a man, they stared at the airstrip and wondered how they could have been taken in to believe that the U2, with all its trailing parachutes, could land safely on this minute piece of cleared Bush.

Cold beers or not, Erdi had his comeuppance. Later that year, he was briefed to give the American ambassador a close-up view of the operation and its potential. His camp administrator, Nigel, an old Etonian, was setting up the final arrangements for the welcome and lunch. For security reasons, Nigel had ordered that no deliveries were to be allowed to enter the compound for the duration of the visit. Erdi and Nigel waited. At first nobody was concerned by the delay, since that was the nature of things in Nigeria, but as evening approached, Nigel felt he should make enquiries. At the gate, the visitors' book was open: just one potential interloper had been turned away. He was, the guard said, a commercial traveller from the United Africa Company—Unilever's subsidiary known locally by its acronym UAC. Slowly it dawned on our old Etonian what had happened: the ambassador's driver had announced 'USA', the security guard had heard 'UAC' and sent him packing. It says little for the penetrative powers of the visitor, but much for the decisiveness of the guard. For Erdi, it was mission unaccomplished.

A few months after Red Adair's first visit, we thought we might need to call on his powerful juju again. We were drilling a well about three miles from camp on dry land in the Bush. Wells are at risk when you breach the first structure if gas is close to the surface and you cannot get enough heavy mud into the hole to control it. When this one reached that stage, we set about clearing nearby villages, since if things went wrong the whole area could be enveloped in the explosive mixture. But curiosity soon drew villagers back to stare at this well as it vomited huge chunks of rock high in the air on a cloud of gas. The gas had formed into a white mist and we could see it drifting towards a nearby flare. We were convinced there was going to be a mighty explosion—but to our amazement the cloud passed through the flame and simply snuffed it out. Whether the temperature of the gas in its grey frozen state did the trick did not matter, for the well was calming down, the gusher had reduced to a gentle controllable emission and the gas had dissipated itself on the wind.

The organization was growing stronger by the day as our budgets grew, and the area of our operation spread from the Eastern Region into the Midwestern Region. As it grows, an oilfield attracts more talent to manage the huge investments involved. Nigeria was the beneficiary of our experience in Venezuela, and technical competence arrived in abundance. A pipeline is a pipeline no matter where it is, and a drilling rig is the same whether it be on the banks of the Niger or Lake Maracaibo. But the human and political problems and environments are always different.

Men and women arrived with their varied backgrounds and experience, their own judgments and prejudices. These inevitably complicated matters, particularly if they were magnified by communication failures. A managing director arrived from the Far East who was accustomed to fresh flowers on his desk every day, and everything polished and in its proper place. As soon as he saw his office he called the administration manager and explained that he wanted a total redecoration. The administrator, an experienced Nigerian, said that if that was what he really wanted, he would put it in hand. The managing director insisted on instructing the contractor himself. He took pains to explain that the walls were to be repapered and that the chosen design was to be hung with great care. There should be no gaps at the joins or edges or anywhere else. Work duly began on Friday night and by Saturday it was completed. On Monday, looking forward to seeing his fresh new office, our MD strode down the corridor ... and opened the door to a steam bath. All the air-conditioning vents had been meticulously wallpapered over—no gaps, as instructed—and in the last 24 hours the paper had begun to unstick in the heat. Strips hanging limply from each vent flapped a desultory greeting. As Burns wrote, 'The best laid schemes o' mice an' men gang aft agley.'

Embarrassed but undeterred, the MD turned his attention to the carpet. He wanted the rather utilitarian carpet tiles to be replaced by a luxurious Axminster. When it arrived he briefed the contractor again. The carpet was a replacement, he was told. It had to cover the exact square footage of the existing one, no more and no less. The contractor nodded assent and set to work on Friday evening. Our man arrived on Monday, excited to see his new carpet, and there it was, an exact replica. The beautiful Axminster had been cut carefully into squares and laid in tiles precisely as its predecessor had been. In shocked amazement he called for the administrator, who shook his head sagely and suggested the MD sit in on our supervision course. Leaving instructions and walking away, he pointed out, is a recipe for disaster. Burns was right.

Throughout the operation lessons were being learned on both sides of the cultural divide. In my first job we worked in a long warehouse building with a main corridor that ran through the centre, offices on either side. The door was policed by Peter, a small but upright elderly man. On Remembrance days he wore his ribbons, which included a Burma star. He had fought with the West African Rifles in that gruelling campaign and had been duly decorated and commended.

The office messenger, Njoku, delivered mail and made coffee. The Dutch head of department, Bob Seelt, could not function without his coffee and on that day it had not arrived. He called to Njoku who, to Bob's bafflement, walked straight past the door paying no heed. An intense thinker who could quickly make a simple problem complex, Bob called his advisers. Was this slight the first indication of personal independence? Had Njoku fallen for Indonesia's anti-Dutch propaganda? As the possible explanations flowed into Bob's fertile mind, our employee relations adviser arrived. A former district officer of many years' standing, he asked simply 'What do you want?'

'My coffee,' came the sardonic reply. This was heading into realms of strike and insurrection.

'Eugene?' the adviser called, mildly. This was a name we had not heard before in the building. Njoku appeared. 'Coffee, please, for all of us.' The adviser explained that we had witnessed a good example of effective communication. Eugene had become a Christian the day before and answered only to his new Christian name, which we now all knew to use if we wanted our coffee!

The workshops head was an experienced, well-travelled Dutchman. Like his Australian opposite number in the machine-shop, he had seen wartime activity; the Dutchman under occupation and the Australian on attack. They came to me together with a problem that involved dissent among the artisans. But their complaint seemed not to have a focus. The story was extracted rather than presented—it was not about money and it was not about gambling—but still I did not really understand why they had come. I promised to investigate.

I put one of my best industrial relations men on it and in two weeks he came back with a bizarre story. A very small man, a first-class instrument technician, was the source of the trouble. He was the juju man, and he walked unchallenged through the shops, dispensing judgment on a wide range of social issues. I decided to take a look at this fellow and there he was, installed in an air-conditioned room which he had set up like a laboratory, accentuating his aura of technical genius. He carried out his instrument work with meticulous tidiness and everything was completed and labelled. The man had skilful hands and weasel eyes. To suggest he looked demonic would be unfair, but it would not be far off the mark.

We confronted the two managers. 'You didn't tell us the whole story. You have to do something if you want your shop to settle down. This man is like a fox eyeing the chickens—and the chickens ain't happy.' Normally these hardened oilmen, seasoned professionals from opposite sides of the world, would act first and think about it afterwards. Normally we would have to restrain them from hasty action and encourage them to consider the consequences … but not this time. Getting to the point I asked them if they believed in juju. Both said yes, and both said they would not touch it. They knew it was the man who had been given the exclusive benefit of air-conditioning who was causing the problem. His perceived invincibility had been reinforced by this preferential treatment. However, after the industrial relations man and I had paid our visit, which he did not react to but was certainly aware of, his disruptive activities subsided and morale on the floor improved. He stayed where he was, but perhaps sensed that the tigers were now watching the fox. We had given him the message. The oddness of one-handedness had its advantages: here was a man who had tangled with something bad and survived. He too had strong juju!

*   *   *   *   *

The Galore Palace was a drinking spot, a ramshackle hut with a corrugated-iron roof to shelter customers from the rain and duck-boards to keep their feet dry. Its music drowned out conversation from 20 yards away and there were usually ten or 15 carousers reeling away on the mud floor, miraculously avoiding the serious drinkers. There was ice-cold beer and noise and you could be sure there was always something going on.

One day a man arrived with an intriguing contraption that had its own embroidered covering, an ornate letterbox on the front and a tray at the back. Inside, you could make out wheels and spindles that turned with the handle. This was a money-making machine. Put in ten naira and, after a short wait, out came 20. The operator demonstrated by metamorphosing his own notes. One timid customer tried it out, and then another before the machine had to be carried out to the van for maintenance. The next week the number of customers grew, but again the machine got stuck with money inside. Out it went again to the van for maintenance, before coming back to deliver the notes. The next time the man arrived it was payday, and customers eager to double their money poured in, as did the cash. The machine stuck and as usual was wheeled out for repair and maintenance—but this time it did not come back. Alcohol and excitement had numbed the senses, and wages disappeared along with the alchemist.

In the early '60s the Royal Shakespeare Company ventured to Port Harcourt and staged *Macbeth* in the open air. A huge audience sat on the grass, totally engrossed, reacting loudly to Lady Macbeth's evil machinations. The narrative was well understood and acutely felt, the violence and lust after power—brilliantly portrayed—a chilling theatrical foretaste of what this new country had in store.

Regional independence was under way and we celebrated with fireworks. Our accident prevention department, under the leadership of a cadaverous Mexican, managed the display. It was spectacular, and best of all was the final, massive rocket. The pièce de résistance was lit with great showmanship before it careered off horizontally to bury itself in the guesthouse roof, which duly burst into flames. The villagers roared their appreciation: this was a triumph, but the best was yet to come. The department's fire wagons responded immediately, speeding across the open lawn and dousing the guesthouse with powerful water jets. Diners who had minutes earlier been enjoying peace and quiet away from the display found themselves under fire from water cannon. In Nigeria, they learned, action was everything, and they would be much safer being part of it.

The buskin boot of *Macbeth* stimulated local schools to take to the boards, and our nearest school put on *Dr Faustus*. We sat on benches in a dimly lit classroom watching a faultless performance. The doctor, a small boy in shorts with a mortar board, discharged his role brilliantly, but it was the seven deadly sins who prompted maximum audience participation. At one point one of the raised benches fell over, toppling a mother and baby onto the spectators below. The sixth-form-sinners were hissed and catcalled—Greed took the brunt but Sloth also elicited a spirited response—and were awarded the lion's share of the plaudits despite the miniature doctor's magnificent performance. We went home by candlelight remarking how delighted Marlowe would have been with the production.

*   *   *   *   *

Union negotiations were coming up and things were unsettled. I was to make the preparations and assist Hez, an Ibo from Owerri who was a member of the international labour organization and an industrial relations expert. The collective agreement was all about wage levels, gradings, overtime rates, allowances and vacation entitlements. At least that was what we thought. The demands that came in ranged far outside our list and included pensions, retirement gratuities, medical treatment, transport provision, bad-weather compensation and more.

The first session started slowly. Hez, rather like Peter Tingom in the longhouse, started by asking for the union's statement, which contained most of the demands. He asked for confirmation of its completeness and then began to take the items one by one, restating and reshaping them and finally reaching agreement on what they were asking. This was masterly. We had not shown our hand and yet he had already compressed the list.

Hez's next move was to group the demands: wages with allowances, holidays with hours of work, overtime with travel time and so on. As he grouped them, he moved some into a category designated 'for further study'. We were on day three before we put our offer on the table, and by that time the workforce was impatient for results. The attraction of winding it all up by the weekend was too great for the union and we secured our first collective contract, uncomplicated and setting no awkward precedents. Well done Hez. For me, always impatient, it was a wonderful lesson in how to negotiate. My careful preparation had been valuable—but only because Hez had used it so well. He listened, negotiated, listened again and restated until he got the answers he wanted; above all he was patient.

However, welcome as the contract was, its success was short-lived. A breakaway union was emerging, led by a man called Constant who was bad-mouthing the incumbents. This was a messy situation and Nigeria was rich in lawyers who knew how to make things messier. The dispute made its much-publicized way to the High Court in Lagos, where our agreement was overturned and the breakaway union applied for recognition, insisting there should be no delay. Hez was very disappointed, but when the ballot was held the new union had the numbers. In a typically astute move, MD Stan Gray personally welcomed Constant into the boardroom. It had been a frustrating time but we made the best of something we could not influence. Constant, by nature as well as by name, was eager to get a deal so we signed up, and peace was restored. The breakaway had not disturbed the operation, but it was bad for morale to have infighting and tension within the ranks.

Port Harcourt now had a large industrial area. The Michelin tyre company and Alcan the aluminium company were there alongside a range of oilfield contractors. Each one was beginning to have union problems. New unionists were appearing, advertising their radicalism and professionalism. It was time to take a look at this. Getting in among them was not going to be easy, so I took advice.

The Reverend Michael Mann, the vicar of the Anglican Church in Port Harcourt (and future Queen's Chaplain at Windsor) set up a study group to which he invited key unionists. I was keen to be involved. Only about ten turned up, six of whom were would-be activists and at least one of whom was in the local MI5's books. He had been in Vietnam and had seen active service with the Vietcong, so his arms and explosives competence were high—but his union proficiency was sadly non-existent. The men knew only about his lack of performance, so he was never appointed. The meetings were tricky, but the involvement was an education for both sides.

Before 1960, Port Harcourt had been a relatively quiet town with trading stations and warehouses alongside the docks and the harbour. The residential areas were generously laid out and the 1930s-style recreational club had a beautifully designed golf course, tucked in between the homes and the creek. The railway station was close to the port and ran through the outer reaches of the town. By the end of 1960, as the oil industry moved in, the area entered a period of explosive growth. Roads became choked, and shanty accommodation spread like a weed, engulfing what had now become a city. The shanty town had all the depressing features of slums without any infrastructure. Many hopefuls arrived from outlying areas only to slip through phases of unemployment into desperation and then self-destruction. Some more robust individuals turned to crime; others gave way to hopelessness. This social disaster hardly registered with the authorities and was not addressed in any way. At least in Scotland the slum clearances drove the dispossessed to America, where there could be new hope. From Port Harcourt, there was no America to escape to.

Jo's stint as a teacher in Brunei had made an appreciable difference and been well received locally; it had also strengthened her determination to be involved in and contribute to the communities we found ourselves in. She became increasingly concerned about the health of the growing number of unmarried mothers and their children in the shanty towns and set to work rallying other women, both Nigerians and expats, to join her in helping to relieve the growing problem of unwanted babies. Some were orphans, many were arriving as a result of prostitution. Using social connections and other resourceful initiatives, she and the group raised enough funds to create a home for at least some of these children and give them a better start in life.

The home, set up on the edge of Diobu, was not without its initial hiccups. Health inspectors pointed out that they were adding to the accumulating sewage problem. Their nappy-washing poured effluent on top of an overloaded drainage system that was already a rats' paradise. The solution lay in a deal with the night soil collectors who agreed, for a small additional payment, to deal with the extra waste.

Financing came from many sources, even from the local madam, who saw Jo as a redeemer, ready to forgive her sins as long as she paid up. Jo visited regularly to collect. On one occasion she took a Shell head office visitor who was keen to see how this example of community involvement worked. Jo introduced her to the madam who, assuming the visit signified collection day, unwound her sarong to reveal a large sporran hanging between her bare legs. She plunged in her hand, retrieved a fistful of notes and duly handed them to Jo. Turning to our visitor she said with a gleam in her eye, 'Anything else I can do for you?' Jo hastily shepherded him out of the door and past the sign that read in large painted letters: PLEASE TRY NOT TO PISS HERE.

*　　*　　*　　*　　*

Employee relations were part of my expanded portfolio, and things were not looking good. Pressure had been growing to address what was loosely described as the expatriate differential. This concerned the difference between the salaries paid to expatriates and those paid to Nigerians for doing the same work. The problem was exacerbated by the fact that expatriates lived in excellent purpose-built accommodation, while most of the Nigerians had to find housing in the expensive, inadequate Port Harcourt market. Part of the differential was accounted for by the extra expense expatriates might encounter working outside their own country. Most people accepted this, but there were still rabble-rousers who enjoyed making a divisive issue of it.

Insecurity among our Nigerian staff was growing, not just because of the deteriorating political situation, but also because the terms and conditions Shell offered were far better than those found anywhere in the job market. To lose a job with such benefits would be a huge blow, so there was a strong incentive to defend, protect and advance staff interests on a broad front.

Ted, my boss, took the unusual step of appointing Alex, the president of the Staff Association, as his personal assistant. This might be—and was—seen as a deeply cynical move, but in fact it was a master-stroke. It gave recognition to the association without its having been sought, and gave Ted a spokesperson within the association who would understand the company's position. The strategy also gave us the time and opportunity to launch a home-ownership scheme, which would solve an important element of the differential while bringing stability to the staff. Owning a house would improve staff morale and, because it was through the company scheme, it would boost confidence in the idea that we were here to stay. We bought land and organized house-building to get the scheme off the ground, which did much to diffuse the tension. In time the scheme and the loans were taken over by the bank, and so became part of the financial infrastructure.

Alex and his wife, Beatrice, asked Jo and me to lunch at their home near Arochukwu, a principal historic town in Ibo land. As we drove down through rich green countryside where yams, corn and avocado trees grew in profusion, we agreed how lucky we were to live in such a vibrant and fascinating country. 'I love this place, its people and its impossible problems!' Jo said. She was by then teaching at a local school and had begun writing a Nigerian history textbook with a friend, Eunice Nwankwo. Because of her historical studies she was as close to understanding the growing political pressures as any of us—probably closer.

The visit turned out to be a great deal more than we were expecting. A curry lunch in rural surroundings is always something to look forward to, but this was not the focus, nor was there a business or investment proposition in prospect. What we were given, to Jo's total delight, was a glimpse of history. When we had finished our meal, Alex said there was something they wanted to show us, and asked us followed them into the Bush.

After a mile so we stopped and he said, 'This is why Arochukwu was well known and feared. My ancestors were the guardians of this place.' We were intrigued. We were standing above a fast-flowing stream that cut through the fertile slopes. On the left was a path, barely discernible in the undergrowth, leading into a cave or tunnel. The gully was strangely disturbing; the rich, open greenery contrasting markedly with the half-hidden darkness beyond. Alex explained that the place was where families brought their young to face either blessing or sacrifice. Children were offered up and sent down into the gully to their fate; if they returned, they were blessed by priests and would in time enrich their village, but if the water turned red and they didn't return, then they had been chosen for sacrifice, a sacrifice that would bring both honour and great riches to their people.

The grotto apparently had a long tunnel through which the 'sacrificed' were taken and handed to the black slave traders. From there they went to the port of Bonny, and on to America. On the way back to Port Harcourt, Jo was ecstatic. Having been a student of American and Nigerian history, having researched the slave trade and the emancipation won in the bloodiest of civil wars under Lincoln, she felt it an enormous privilege to have seen how and where this had happened. No favour was sought in return. This was not about Shell; it was about Jo and her teaching and the baby home and all the work and compassion that had built her into Ibo society. This was Beatrice saying, 'Let's share something with this good woman that she will understand and appreciate.' And she did.

*   *   *   *   *

One of our young men working in personnel, a talented footballer, was killed in a car accident. Paul's new Peugeot 404, a symbol of his success, was written off. As I sat in the church in Onitsha with his many relatives and friends, my mind absorbed the measure of the tragedy. His achievement in getting a job in the oil industry had changed his status completely—though ironically it was through that prized fast car that he had lost everything. This thought was confirmed that night as we had supper with a Cabinet minister who described the great hopes they had had for Paul. Chinua Achebe describes in *Things Fall Apart* the burden put on those emerging into a new and more prosperous life.

Driving home afterwards on a wet, potholed road, I thought about my own life. The environment in Nigeria suited me well: there was endless work and challenge. There was a company ethos that rewarded initiative and application, and we were in a country full of aspiration and optimism. It all fitted; almost it seemed by cosmic design.

When I was 11 years old I was commended for an essay I had written— an achievement commonplace for my brother, but unusual for me. My essay was about meeting a large black man on the shore near Kirkcaldy, where he was looking out to sea. I had asked him what he was looking for. 'Home,' he said, and in case I thought he meant North Berwick, he described the villages, the jungle, the heat and the darkness of the night with all its noises of busy insects and animals crashing through the Bush. I knew well the story of Mary the mill-girl missionary from Dundee, and most of the colour in that essay came from there. The commendation, I suppose, made it memorable, but I still do not know what inspired me to write it. The idea of Africa and its excitement must somehow have been planted early. Jan's almost-forgotten astrological charts swam back into my mind that night.

Sir Francis Ibiam, a graduate of St Andrews, tells a story of how, when he arrived from Nigeria, he stood looking around the station and wondering what sort of place he had come to. He was approached by an elderly gentleman who asked him where he was going. When he replied, 'To St Andrews University to study divinity,' the old gentleman became agitated and cried, 'No, it must be medicine! Do the body first, and when that is done attend to the spirit; it has more time—infinite time!' Sir Francis puzzled over this as the train made its way slowly across the Forth Bridge and on through Fife. By the time he reached St Andrews his mind was made up: medicine it would be. Perhaps Sir Francis was the man on my seashore.

*   *   *   *   *

This assignment to Nigeria meant one history lesson followed by many history lessons as Jo collaborated on her textbook. She and Eunice told the story of Lugard who, marked by his East African experience, established that expatriates had no land rights, thus avoiding the tragedy that subsequently played out in Zimbabwe and to a lesser extent the disruption of Asian settlements in Kenya and Uganda.

Because of the policy of indirect rule, Nigerian politics was an exercise in self-development. The North stayed autocratic in its governance, disciplined in its habits, tolerant of the occupier and devout in its religion. It looked towards the desert, a hostile environment but peaceful, being so little populated; and to the South, where disruptive ideas and forces were at work that could undermine stability. Already the Northerners had seen commercial acumen in their midst as the Ibo monopolized transport and expanded their own trading businesses. The Ibo had no religion in common, unlike the Yoruba. The Yoruba had assimilated the Muslim faith in the northern areas of their kingdoms, and found comfort in the underpinning of regal structures. Chiefdoms in this environment were not threatened by the democratic protestations of those further south. The oratory of Governor General Nmadi Azikwe, who came to be known as the Father of Nationalism, fell on deaf ears, except when it touched the issue of independence.

'Show the light and the people will find the way,' was the clarion call on Dr Azikwe's newspaper *The West African Pilot*. The drive to independence had no resistance, only support—Macmillan's 'Wind of Change' speech had made sure of that.

By deft manoeuvring, Ahmadu Bello in the North, Michael Okpara in the East and Akintola in the West had taken power, and eventually in 1960

Tafawa Balewa and Festus Okotie Eboh, from the Midwestern Region, emerged respectively as prime minister and finance minister. So the scene was set for 1 October 1960.

Since our arrival we had set up home and begun to build a network of friends. This was not difficult; Nigerians are naturally gregarious and hospitable. Political discussions were robust; convictions strongly held. Socialism melded easily if not logically with personal advancement and wealth, Catholicism and Protestantism mixed without undue rancour and there was complete ignorance of the Muslim faith. As it would turn out, the Catholic descendants of Cardinal Wolsey were much closer to politics than their Protestant brethren were.

On the eve of Independence Day a frigate had arrived in Port Harcourt, either to join in the celebrations or perhaps to ensure the safety of the large expatriate community. In any event we were advised not to venture out that evening. However, Hez Ofurum had invited us to a party in downtown Port Harcourt, and there was no way we could refuse. His parties were the best. There were toasts and cheers at midnight, and dancing followed into the early hours. As we made our way home through the villages that surround the city, we expected to see riotous drinking and dancing—but we saw no one. Everyone was asleep, as if independence was for someone else. And that, unfortunately, was the way it soon turned out.

Within the company, the day-to-day operations were unaffected. There were no incidents of insubordination or ill-discipline, or even a change of attitude among our people. It was business as usual. My role grew as the company grew and I was now responsible for training as well as industrial and employee relations. I had read a report published by the Ford Foundation, an American organization created in 1936 by Edsel and Henry Ford, which had been commissioned by the Western Region government to tackle the problem of petty corruption among junior clerks. The taking of cash for services was endemic and it was undermining the reputation of and respect for government.

Two Ford Foundation executives took jobs in the service at junior and public-interfacing levels. They lived in simple accommodation, cycled to work and learned to accept cash 'incentives'. In no time at all they were just part of the furniture. Understanding the process by which one becomes corrupt is only the beginning; what you then do about it is something else.

When the pair finished their spells as corrupt clerks they set about designing material to tackle the problem. It was all about self-esteem and respect for the job you were given by your elected government. Their training method was fascinating. It was called T Group, and it worked on the basis of simple questions and openness, and waiting for your group to come to you. This could be very painful but it was a necessary step before you could build a solution. Somebody in New York had apparently thrown himself out of a window as the new technique of 'silence and waiting' took its toll. This did not worry me too much—after all, our training centre was on the ground floor.

We hired them. George our Ford man went through our objectives: to build a sound relationship across the racial divide; to formulate a process by which the executive could isolate, identify and resolve problems that had racial overtones, and to create a positive working environment where trust and confidence in each other were implicit. George knew the ways of West Africa, having worked in Ghana as well as Nigeria. At the end of the first day a Dutch engineer withdrew. The others, in mutinous mood, stayed on. On the second day, Godwin, our training manager, came in. One of the expatriates immediately exclaimed, 'We don't want him in here!'

'Why?' a Nigerian executive retorted, 'Is it because he's black?' A battery of protests followed while George, having asked Godwin to sit down, looked impassively out of the window. When the buzz subsided at last, Godwin said, 'I only came in for my pen,' and strolled out again.

At the end of the third and last day the course members were unanimous in their praise for George and the course. Before he left, he confided in me that Godwin's unplanned intrusion and exit had been a turning point, forcing people to examine their attitude and behaviour. He was tempted to build something similar into his future courses.

We followed up with an executive course based on corporate decision-making. This time Nigerians and expatriates worked together, dealing with specific investment and operational decisions that had immediate results and potentially problematic consequences. The decisions involved unions, discipline, fraud and local community issues, and the idea was to broaden the experience of those who had specific responsibilities so they could see how their contribution worked in the context of the overall operation. People with technical skills needed to understand how their decision-making impacted revenue and cash flows, for example, and financial administrators needed to see how their systems affected the teams on the ground.

By far the best-qualified group was led by the Scottish finance director, a gold-medal accountant whose team completed the investment cases with ease. But while they focused on accounting issues, they left other potentially damaging matters untended. When the local community complained about land acquisition and crop damage, for example, the team ignored them, as they did the follow-up letter from the local MP. The next stage was a letter from the minister, to which they replied in forthright manner telling him he was wrong and the matter was closed. By this time their cavalier approach had also brought union involvement to a critical head, and eventually both issues required a considerable investment of time and money to resolve.

The director was mortified, and protested that their financial answers had been unimpeachable. We issued a stark message in reply: listen and pay careful attention to those on whose ground you conduct your business. Listening is the first step in understanding. The Nigerians had an excellent practice: they repeated what they had heard to ensure they understood it correctly. It was a habit I began to copy.

This hugely instructive course was repeated several times with as much enjoyment for the participants as for the organizers. The message that failure to listen is not only arrogant but destructive was dramatically brought home when BP was expropriated. Paying no heed to the Nigerian embargo on selling crude oil to South Africa, BP went ahead and sought approval from the British government to sell on to the country. Nigeria responded by taking over the company's remaining stake in Nigerian oil operations, and so we lost our partner.

Repeating my Brunei experience, we launched the Shell Scholarship Scheme—20 scholarships a year to the best engineering universities in the UK. This was a success and in time produced first-class honours degrees every year. As part of the scheme I visited the universities at Ife and Ibadan, and was struck by the strength of their facilities and by their cosmopolitan approach. It seemed strange that we, with our village full of PhDs and MScs in earth sciences and engineering, should have no contact with these intellectual centres. To remedy the situation we planned a visit with our top scientists and engineers, and it was a huge success. The resulting relationships won us some excellent recruits, but more importantly the initiative made the well-educated people on our teams take an interest in the country they were living and working in. Recognizing you have something to learn and understand is a giant step towards building a community, and it is the strength and empathy of that community that safeguards your people and your investment. Societal involvement, training courses, sporting activities and intellectual exchanges are all part of the same thing.

In the oil camp, points of cultural difference continued to provoke mystified, sometimes uncomprehending, reactions. The St Andrew's night celebration, for example, when Beatrice, who was also the leading Nigerian contractor, boiled up 300 haggises in a massive cauldron. Many popped open and needed to be rescued, others made it to the plate intact. They were served with mashed potato and yam, the local alternative to neeps, but by the end of the evening they were all uneaten except for those that had been enjoyed by a few Scots. Beatrice surveyed the untouched plates and shook her head, utterly baffled. Had she not followed the recipe? If so why did nobody complain? These were strange peoples!

Christmas in the oilfields was preceded by Sinterklaas, the Dutch festival. Nigerians understandably found this weird: Sinterklaas's assistant, 'Black Peter', was traditionally played by a recognizable member of the Dutch community with a blackened face and arms. One of our Nigerian friends commented that with 80 million candidates to choose from, this job should really be Nigerianized.

Christmas itself has its weirdness, of course. In our version the white-bearded (so clearly elderly) Santa Claus was helped and cajoled to dish out presents to bemused youngsters before being dragged away over the gravel on his sleigh. It's difficult to say whether the absence of chimneys, snow and reindeer made it more, or less, bewildering.

When respective national days were celebrated, the British gathering was always low key, albeit with outstanding cabarets filled with barbed quips about the idiosyncrasies of our European cousins. Dutch nights were a wonder of construction: windmills, barges, coffee bars with unlimited Genever and loud music. October 1st's Independence Day celebration—Nigerian night—was eagerly awaited. When the first anniversary arrived there was no sign of any preparation; no construction, no rehearsal, no cooking. This, the cynics said, was typical. They do not buy into celebration and they would not share it.

At four o'clock, however, a lorry appeared. It was loaded with huge palm fronds which were used to garland the wooden pillars around the dance floor, making it look like a clearing in the Bush. Then large half-drums appeared, filled with sticks and coal, and soon six barbecues were alight, chicken and lamb sizzling. From nowhere appeared cars filled with musicians. They hooked up their electric organs and guitars and by six o'clock were tuning up. Bars were opening with ice-cold beer and red wine, bottles of smuggled gin and whisky appeared and we were almost ready. At 6.45 p.m. the reception committee arrived, the women resplendent in extravagantly coloured wraps and magnificent head ties, their escorts impeccably turned out in white dinner jackets.

The band struck up and the party began. The beat was irresistible and all the guests were on the floor. The Nigerians know how to throw a party. That night we were all family, work was forgotten, rank was irrelevant (with the Aussies it always was!) and we had ourselves a ball. The best dancer on the floor was the unrecognized non-member, rumoured by some to be a visiting Cabinet minister who had just dropped in. He was in fact our own jolly barber who, in competition with his rival the Praying Mantis (named for the creepiness of his fingers around your head) toured the housing area cutting hair. The Mantis did not appear. Either he did not have evening dress, or we did not recognize him among the crowd.

*   *   *   *   *

Before independence, there was just one university in the country, the University of Ibadan. An influential commission had recommended adding new universities to each of Nigeria's three regions and the capital, Lagos. The law establishing the Northern Region's new Ahmadu Bello University was passed by the legislature in 1961. It was named after the region's first and only premier. Shortly after independence, our MD Stan Gray was invited to the Emir of Zaria's palace to discuss the new university, and Stan asked me to go with him. The palace was made of mud baked into solid brick, and the walls were extraordinarily thick, as was the roof. As we were led in by the emir's entourage the temperature dropped, and it continued to drop as we went further, passing over beautiful carpets into the inner sanctum. The bejewelled emir, wearing a long white gown, greeted us warmly from his throne and asked us to sit down. I was interested to notice that he had six fingers on each hand.

Having discussed our scholarship programme, we talked about Shell's activities in the oilfields, though Stan explained that we did not have high expectations of finding oil in the North. The emir took the news philosophically, seeming satisfied that his wealth would come from groundnuts and cotton. (Had we known then that 50 years on there would be rich gas fields in the Algerian desert, we might have painted a different picture!)

After this discussion and a cup of strong coffee we all made our way out into the heat and the paddocks where the emir kept his polo ponies. These beautiful animals recognized him immediately. This was the old

North which Lugard had left untouched. His policy of indirect rule left in place structures and institutions which, in the years that followed, became increasingly archaic until revolution severely dented the aged architecture. This was only a couple of years away, but on that day we had no suspicion that the eruption was so close.

On the way home we reflected on the vast difference, seemingly in light years, between the political structure and the educational penetration in North and South Nigeria. At the time we felt we could define the East reasonably clearly. The Ibos put a premium on education in schools which, as in Scotland, were actively supported by parents and even more actively supported by the Church. All denominations were involved, but the Catholic brotherhood was by far the strongest.

In the land to the east, neighbouring Cameroon, lay Calabar, the town of Mary Slessor, a jute-mill girl from Dundee who became a missionary. Having learned Efik, the local language, this red-headed slip of a woman gained the trust and acceptance of the locals and started teaching Christianity. She also promoted women's rights and protected children, especially from harmful superstitions such as the one which, until then, had routinely led to the infanticide of twins.

The Efiks now were a quiet, well-educated people, religious and with a relatively calm political ambition. To them the palm oil price was more important than the oil price. The fire of political debate was far less evident there than it was in the Ibo heartland. To the south through Ogoniland, which was to become more troublesome as time wore on, their land became the battleground of pirates and plunderers of the delta oilfields. The Ogoni live in the marshes and mangrove swamps, areas where virus and mosquito seek out the living and punish them.

As we tackled our international relationships, bringing higher levels of understanding and respect among colleagues, political alliances were beginning to show signs of strain. The idealism and ambition of Independence Year was ebbing away. We worked obsessively and introspectively on the problems of our fast-growing company, though a few of us became increasingly concerned about the changing political environment. Federal independence was supposed to be the culmination of the movement towards political maturity and freedom; it came and went, but little changed. As we saw it, the federal structure put in place that year was an attempt to balance the main power blocks that made up the nation: Hausa and Fulani in the North, Yoruba in the West and Ibo in the East. These tribal blocks dominated the regional assemblies and controlled the produce marketing boards that paid low prices to farmers and built up surpluses. These in turn were directed to large investment projects. The location of the projects was more about local political advantage than logistical or economic benefit, which created tension and bitter feuding at a regional level, and it was only a matter of time before its corrosive effect would unsettle national politics.

Division of the federal parliament seats gave the Northern Region constituencies 50 per cent representation and the Eastern and Western Regions 25 per cent each. The army was set on the same principle numerically, but it was only in 1961 that there was a move to make this division applicable to the officer cadre. Until then the army was predominantly Southern, so the move was deeply divisive.

The political parties in 1960 were the Northern Party Congress (NPC) in the North, the Action Group in the West and Azikwe's National Convention of Nigerian Citizens (NCNC) in the East. The NPC had made its alliance with the NCNC leaving the Action Group out in the cold. By 1962 its leader, Awolowo, had dissension in his own ranks and had broken with Akintola, whom he had tried to remove from the premiership. This was becoming ugly. We sat in the East as uninvolved spectators. The prime minister, true to his relationship with Akintola, declared a state of emergency and suspended the Action Group for six months. Akintola returned as premier in the Western Region, this time in alliance with the NPC.

The state of emergency in the West was bad enough, but it was exacerbated by a corruption charge brought against Awolowo and the Action Group. It was alleged that they had bought out large car dealerships, paying handsomely for them, and then allowed the dealers to buy them back at bargain-basement prices. All were happy—except the taxpayer and the opposition, who might not have done the same, or might not have thought of doing it—but anyway did not get the chance. Awolowo went down, as did others. The dealers remained intact, no doubt waiting for the next move. They had no party affiliation, only the perfect motive: profit!

The Action Group was now in deep trouble but their fortunes plummeted further in 1963 when the Midwestern Region was carved out of their heartland and a major part of their support disappeared over another border. The blatant misuse of power during a sensitive time and the ruthless nature of its execution left many ordinary citizens aghast, and the non-political pillars of society—the Church and the army— disenchanted to the point of mutiny.

A national census was taken in 1962 to authenticate the old formula for power (50:25:25), and the result suggested there had been a 70 per cent increase in the East/West populations since the last census in 1958, and 30 per cent in the North. We stared at the indelible stamps on our arms, applied at the polling booth to prevent people from casting multiple votes, and then at the results. We were assured by friends in the East that in 1958 nobody had counted the swamp and river populations properly. Having flown over the swamps and motored through the mangroves, I doubted this would have made much difference. A recount was ordered and no sooner had the old stamp washed off than a new one was applied. This time the Northerners found eight million previously uncounted people. Apparently Hausa and Fulani do not like to be counted and neither do their children, but this time they had been persuaded. This was a fantasy world, explanations each time getting further from the truth.

With the status quo unchanged and colonial rule of thumb still prevailing, the scene was set for the 1964 election. Alliances were cemented. Akintola's Nigerian National Democratic Party (NNDP) with the NPC made the Nigerian National Alliance (NNA), and they were up against the United Progressive Grand Alliance (UPGA), the old NCNC with the Action Group.

This election brought home how far things had deteriorated across the tribal divide as the manipulation of acceptance papers was followed by stuffed ballot boxes. Manipulation produced 88 unopposed candidates in the North, and in the East 30 per cent of candidates were unopposed. The violence was serious, and Joseph Tarka, head of the United Mid-Belt Congress in the North and allied to the UPGA, was arrested for incitement. Trouble was spreading north from the West.

When the results were in, the NNA had 198 out of the 312 seats but a significant number of seats had been boycotted. The citizens were showing their distaste in large numbers. We were beginning to feel this in our neck of the woods. By now it was the first and only topic discussed around dinner tables with our Nigerian friends. The women were more outspoken, their hopes for independence perhaps having been greater.

Already the word 'secession' kept coming up. The ferment stirring in villages throughout the East was adding to the flood-tide of support for it, and in response Jo repeatedly set out the disastrous consequences of secession in the American Civil War. This was not well received, but it was heard. Walking home late one evening after dinner, one of our closest British friends, Keith McIvor, a strong proponent of secession, was still arguing with Jo. To everyone's amazement—including his wife's—he shouted 'You are Vorster's whore!' I had to explain quickly that Vorster was not a Dutch driller but the president of South Africa. All present bar one collapsed in fits of laughter, until a policeman politely reminded us that it was after midnight. Keith was a historian by education, a high-class rugby player and a totally committed Christian. He left Shell soon after we left Nigeria and became a missionary.

After long negotiation, and once Prime Minister Balewa and Azikwe had signed a pact laying out the principles, Balewa invited the NNA to form a government. It would be a broad-based government including members of UPGA, the seats boycotted would be recontested in March 1965 and the Western Region election would be held in October 1965.

The Western Region election was a bloodbath, manipulation had been rife but the last straw was the decision to make the constituency results a central, not a local, announcement. This, according to cynics, gave time and opportunity for further 'adjustment'. Results were announced 50 to NNDP, 11 to UPGA, 30 to come. Immediately Chief Adebangro, acting leader of the Action Group, announced he was forming an interim government. In short order he and his colleagues were in detention.

Running battles continued for the next six months. As if this was not enough, the government announced a reduction in the price of cocoa, which had been held artificially high during the election period, from £120 to £65 per tonne. It was the beginning of a peasants' revolt. The West was again out of control, but this time in a more serious way. Balewa decided to send in the army.

*   *   *   *   *

The press in Nigeria is free, so news of the events in the West landed on our doorsteps every day. Our focused executives, however, working flat out in a rapidly growing oilfield, took little notice. It was against this background that we had to pull together the human resources needed for an ambitious and demanding programme. Part of my job was to coordinate the manpower requirements. My clients were the department heads, and my contractors the Shell Group in London and the Hague. To many this would seem an ideal arrangement, but in these structures it was not buyer and supplier or go elsewhere; it was more complicated than that. To me, selecting and retaining the right candidates, the candidates my clients wanted, was the key to our success.

The service companies' agendas were never the same as ours, their range of customers and priorities being very different. Here I was at an advantage, since my clients knew not only what they wanted but also who they wanted. The depth of expertise within the Group was amazing. Geologists and geophysicists, scientists and palaeontologists were high achievers drawn from top universities in Britain and Europe. Their and our success was directly related to where they chose to drill. The Swiss, under Erdi Frankl, were particularly successful in Nigeria.

As a young man, Erdi had been an Arctic explorer. He had learned how to keep his dogs alive in sub-zero temperatures, protecting their lungs with careful pacing, and protecting his own body by strictly limiting exposure. His description of warm blood returning to his sacred member was excruciating. Erdi always travelled with his own outboard motor, which he knew down to the last nut and bolt. This real explorer would never get stuck up the Amazon or in the Niger Delta.

He and his team, with the aid of 3-D seismic data, had success after success in the delta. The drillers were predominantly Dutch, led by the impressive Arie Van't Hof. These were the true heroes, working in the swamps with high-pressure wells, making haste slowly, and deciding when to weight up the mud.

The engineers, whose work was coordinated through Port Harcourt, built the pipelines and block stations that increasingly fed the oil terminal. One of the most outstanding people I came across in my working life was engineer Lo van Wachem. Lo and his father had been caught by the Japanese in Indonesia, and had spent five years as prisoners of war. Lo's father had educated him so well in the camp that after release he took his degree early. By the time he came to Nigeria he had already worked in Venezuela.

Slim and physically uncoordinated when it came to ball games, Lo was unassuming but extremely bright. Equally skilled at engineering and staff management, he rose quickly through the organization. He was not motivated by money or plaudits; his goals were achievement and friendship. As a neighbour he made good the shortcomings of a one-handed husband—hanging pictures, mending taps, fixing windows ... all things which, untended, irritate a wife! Later, after two decades roving from Venezuela to Nigeria to Brunei, Lo returned to The Hague headquarters to run worldwide exploration and production.

While the geologists set the locations and the drillers drill them, the petroleum engineers monitor progress and, when the target is reached, begin to measure the reservoir and its physical link to any other reservoir. Mapping the structures and tracking the production trajectory allows the big picture—the prospects and business opportunity—to emerge.

Shell's success has been based on its personnel planning and individual development process, producing the technocrats who could find, develop and build an oilfield and then organize production throughout the world to places where it would be refined and consumed. In Nigeria we had these painstakingly chosen technocrats in abundance.

Discussions in The Hague concerning staff were taken very seriously, particularly for those in the exploration function. The coordinator, van Nierop, to whom everyone deferred, was an elderly Dutchman widely known within the Group. When I first met him he greeted me neither warmly nor coldly; perhaps he was simply exploring. Ted Newland, the administration manager, was to join us at that meeting, but the previous evening he had left me after dinner and gone out. 'Out', for a South American such as Ted, could mean many things, some of which could involve much time and energy—particularly if the cards were good. Tall and lean, he was a languid person, easy to work for and talk to. But he was also a visionary: imaginative and fearless in opinion. He put forward his provocative views in the same way he played poker: testing, probing and at times outrageous. As I began my introduction to van Nierop, it would have been interesting to have had him there.

I outlined to the situation in Nigeria, the possibilities in the Midwestern Region, and the union situation. When the discussion moved on to individuals, he came to life. His knowledge of his staff was encyclopaedic. I pressed on, putting our case for acquiring specific executives in timescales that he dismissed as unrealistic, and so the bargaining began.

We had almost finished when Ted arrived. 'I apologise,' he said, 'but I had business that couldn't wait. I got away as quickly as I could, but I'm only a supporter in this exercise; you have the man who does the business right here.' Van Nierop nodded impassively. Then, almost as we were winding up, Ted said, 'I have one point to make which I don't like and you won't like, but it needs to be said. As our success grows and our operations expand into the Midwest, we should double-bank our gains and build a terminal there.' This was poker with real money—millions! We were way outside the terms of reference of a staff discussion but Ted was right, and he called for my support. I drew out again how things were changing and, in the West particularly, not in a good way.

'You have the most successful streamlined operation the Shell Group has ever seen,' said van Nierop, simply. 'Why should I change that?' He looked around at his acolytes who were, surprisingly, not uniform in their agreement. Ted got up and thanked him and then said,

'You won't have to change it. It will be done for you!'

As I flew back to Nigeria the next day I thought about Ted's attack on the company's strategy delivered in the citadel of technology. His prescient remarks had been prompted by the political upheavals in the West. My colleagues in Brunei had laughed at Barney's predictions of imminent uprising—but in the end he was right. Ted was an oddball, totally unlike Barney in every way—except for his razor-sharp instincts.

*   *   *   *   *

Lo van Wachem had overseen organizational developments that had built an offshore platform in Borneo; now he refocused those skills to develop the fields in the swamps of the delta. Project management was rapidly becoming a new profession.

Lo selected his teams from mechanical, civil and instrument engineers and converted them into oil and gas engineers who could design and build block stations. These would collect the wells' output and move it into the pipeline network. The new organizational concept was headed by a charismatic enthusiast, Hub van Engelshoven, who would later join the board of the Shell Group, an unlikely place for such a restless hands-on type. Individuals were selected for their drive and problem-solving ability as well as their competence and engineering skills. This was the task force that brought the explorers' and drillers' work to fruition, raising exports to a million barrels a day in six years.

The tempo of activity increased throughout the operation. The pipeline contractor found he could not get pipe from the docks to the site in time, so he drove to the Port Harcourt–Aba road with the single objective of selecting drivers. When he was overtaken for the third time by two speeding mammy wagons (light, open-sided trucks), he followed them and flagged them down. He told the drivers what he was looking for. 'Driving pipe,' he said, 'needs concentration and commitment. Concentration to follow the route without delay, and commitment to bring the pipe undamaged to the site. Accident is delay and damage—I want no accidents, no damage, and no delay, but I want my pipe there.' The men were hired, hand-picked like everyone else.

In the middle of this frenetic activity accidents happened, but few more tragic than the one that broke the back of our popular production manager, Eddie Duke. Eddie and his wife, Thelma, were from Calabar, capital of Cross River State. Both were active members of our community who participated wholeheartedly in social events. Eddie, an experienced engineer, was destined for a senior position in the rapidly growing department. His jeep overturned on a narrow village road and hurtled down a steep slope. He was so seriously injured that we arranged to fly him to a specialist UK hospital, but before he could leave his condition deteriorated badly. It was clear a crisis was approaching and our tight-knit community was deeply affected. Three weeks after the day he swerved to avoid a duck, he died.

The funeral arrangements were my responsibility. Eddie was to be buried in Calabar; in the meantime, because of the tropical climate, the body would have to be frozen. This was agreed only after lengthy negotiations with the Calabar elders. They welcomed the plan that Stan Gray would fly down and give the funeral oration at the graveside. The coffin had to remain sealed, since by the time we reached the church, the corpse would be beginning to defrost. The non-Efiks were desperately concerned about this; not only did it go against tradition, but also they did not wish to face any allegations of having tampered with or damaged the body. Tribal nervousness was running high as we set off that Saturday morning to drive to Calabar.

We met in a school hall to prepare for the service and take some refreshments. After an hour and a half, it was clear that things were not running smoothly. My Ibo friends were battling to keep to the coffin-closed agreement. In the end our lead negotiator told them that if we could not agree, we would have to send back the plane with Mr Gray on it, so he would not be able to deliver the oratory. This broke the logjam.

We loaded the coffin onto a station wagon and proceeded through the crowds towards the wooden Presbyterian church. Ten of us lifted the coffin onto our shoulders to take it inside; I was on point right at the front. The coffin was extremely heavy and firmly propelled from behind, and as we tried to deliver it to its place before the pulpit, we took out the door and half of the front flimsy wall. The huge minister had looked unwelcoming as we approached the door; not surprisingly, his expression turned thunderous as we then crashed through it. He preached his sermon: we were all sinners and the same questions would be asked of us as were being asked of our brother Edward. As he roared out his message he fixed us with a glare that must have put the fear of God into his parishioners every week. I had listened to fire and brimstone before in Scotland, but this was on another level.

After a benediction punctuated with dramatic pauses, we made our way through the now-substantial opening with the coffin, and deposited it back in the station wagon. Calabar's steep hills plus the weight of the load were too much for the wagon's clutch, so we ended up having to push. We reckoned there was a good chance that brakes would give way too, so in scrum formation we pushed the wagon uphill, and then chased after it as it threatened to plummet down the other side.

At the cemetery thousands gathered to listen to Stan. He was not usually a natural orator. I wrote his speeches from time to time and he gave them without alteration, but this time it was different and truly memorable. He described his international family. Eddie had been a valued member of that family and his loss touched everyone. There was not a dry eye on that hot afternoon, and every statement was met with a heartfelt cry of 'Good talk!' The image of Stan, standing on that muddy mound preaching with the same intensity as the preacher, left a deep impression. That moment seemed to bring us all together.

*   *   *   *   *

Our absorption with our work and our personal and corporate sadnesses could not by now block out the tragedy that was developing around us on a mega scale. The Western Region debacle was now seen by the public against a background of growing corruption. Corruption starts with the idea that some irregularity is justified by the benefit it can bestow on your homeland and your political base. These 'benefits' were plain to see as you drove through the countryside: new buildings, new plants and even harbours, positioned more often than not for political rather than logistical advantage. But for the agent responsible, something always sticks. It starts like a defensible broker's fee and then becomes a share and then the share grows in Swiss banks.

I have a Nigerian painting by the artist Twins Seven Seven—so-called because he was the only surviving child of the seven sets of twins his mother bore. Inspired by Amos Tutuola's book *The Palm-Wine Drinkard*, the painting portrays a drummer and another man beating time with a seed-filled calabash. Behind them is a mass of branches and leaves, all intermeshed with one another, and as you look more closely you see small faces and then pairs of eyes everywhere. These were the eyes of Nigeria, watching who was taking what. There is no place to hide and your history goes with you and your children. The idea reaches back to the Ibo homeland, Arochukwu, and beyond.

In January 1966, there had been a successful Commonwealth conference in Lagos. At the end of that week Harold Wilson, then the UK prime minister was visiting the East before leaving Lagos to return home. On the morning after he left there was a communication breakdown with the Eastern Region capital, Enugu. This was unusual, and was followed by other strange failures. I was standing in for my boss and joined the management team to review these developments. We agreed to see what we could find out and to revert four hours later.

We had hired an industrial relations supervisor who had financed his studies in New York by washing dishes in the Bronx. It was fitting that this attractive, at times outrageous, Ibo was called Nathan: he was straight out of *Guys & Dolls*. After the morning meeting I called him and his co-supervisors and gave them three hours to find out what was happening. At the appointed hour, the men trooped in. The first three had nothing to say, and then it was Nathan's turn. 'This morning,' he said, 'the Sardauna of Sokoto [Ahmadu Bello] was assassinated in Sokoto. In Lagos, Prime Minister Balewa and Finance Minister Chief Festus Okotie Ebo were assassinated. In Ibadan, Akintola has been killed and in the Midwest and the East the premiers are under house arrest.' Everyone was stunned. I asked where he had got this information, and he said he had bought a box of cold beer and paid a visit to his friend in the airport control room. There he had tuned in and listened to the army wireless network.

At the management meeting, like Nathan, I was last to speak. Nobody before me had any information. Aged 32, I was a decade or so younger than the rest of them and as I looked round I thought, *Nathan, this had better be right or we'll be the laughing stock*. Already doubts had been cast on my growing Nigerian team (of my 100 – x formula, I was the only x). I plunged in and said, 'This must be a military coup. We need to warn our operating units and activate our contingency plans for the camp, as unrest might follow.' The team accepted this, and by six o'clock that evening Nathan was vindicated.

The unrest we feared did not come, but the story began to leak out. Dissatisfaction among military officers had led to five majors, four of whom were Ibos, taking coordinated action to carry out the assassinations. Their grisly work done, they left the military hierarchy to take over. General Ironsi, an Ibo, became head of state and declared that law and order was a priority and essential services must be maintained—so it was back to work and business as usual.

But it was not usual; things would not be the same and there was a strong feeling it would not end here. This had been started in the West but it was now spreading countrywide. The North's shock at the sardauna's violent death had at first prevented the news and its implications from sinking in. Their leader had been killed by Ibos—but the Ibos' leaders had not. Strong words from Ironsi about corruption and tribalism rang hollow, as did his promises to bring about a new constitution and return power to the people. People were not sure they liked the first taste of democracy; some were even heard to say that they did not like independence. Emotions were running high.

Ironsi passed Unification Decree No. 34 to turn Nigeria into a unitary state, and put in place a central directorate where military governors were in charge of the old states. Corruption was to be pursued and tribalism eliminated (as instructed, we duly removed the 'tribe' category from our records). Ironsi appeared to rely only on Ibo advisers, which was a growing concern.

The empire inevitably struck back. Northern NCOs and officers captured and killed Ironsi in Ibadan. Thirty-one-year-old Jack Gowon was appointed in his place and Decree 34 was repealed. Ironsi's death unleashed pogroms in the North and the East, and those who escaped related tales of appalling savagery. The fire had now truly spread and the East was engulfed by outrage. Pogroms in Port Harcourt began and white-robed Hausas began to flee. The police did an excellent job in rescuing them and entraining them to the North. When a murderous crowd converged on our residential camp they released their Alsatians, and the bully-boys fled.

The situation had deteriorated badly, and calming our staff, expatriate and Nigerian alike, was no small task. This is where leadership counts. We had 15 young Yorubas, barely in their 20s, at our training centre in Owerri. We could not leave them there, so my Ibo senior executive, Bethel Njoku, organized for them to be ferried into Port Harcourt in small buses in the middle of the night. Once on the wharf at our transport depot, one of our boats whisked them off to Lagos. It was a nail-biting operation.

As tension mounted, staff from outside the East were taking boats and crossing the Niger towards their home villages. This exodus was depleting our fleet and when I briefed the supply manager, a Dutchman who had been in a POW camp in Indonesia, he looked me straight in the eye and said, 'Bob, this is now a new situation. This is about survival. We can replace boats, but with all our accumulated skill we cannot replace lives. Let them go.' I realised the stakes had changed.

People's personal experience drives their actions and reactions. One of our drillers saw a mob with sticks and knives pursuing a Hausa. Disobeying company advice not to become involved, he waded into the mob yelling 'No!' and a few choice Dutch expletives. This had no effect on the building frenzy, so he grabbed the Hausa by his robe, pushed him into his car and drove off to the sanctuary of the police compound. A life was saved but he was lucky; two lives could have been lost. In the conversation we had afterwards, he explained that he had seen this all before as a boy when he could do nothing to help. A Jewish shopkeeper who had always been friendly to him was being attacked by the Gestapo. He said, 'I felt totally disappointed with myself and vowed if this ever happened again I would not come up short.'

Amidst such tension there are always incidents that bring a new and sometimes amusing dimension. As we reviewed the situation with the ever-calm and resolute Stan, I advised him we had one Yoruba lady still in the camp. She was very beautiful and from a distinguished family, her father holding high office in Lagos. Stan lost his cool when I said she was staying with one of our British staff. He called his Nigerian head of public relations and demanded why on earth amid this mayhem she had not gone back to Lagos. Kanu, a most charming man, shrewd and superb at his job, simply said, 'They are in love.' I thought for a moment that Stan would explode, but suddenly he saw the humour of it all and with a guffaw retorted,

'This I don't need! Send them both to Lagos—now!'

Our little camp in the Midwestern Region was surrounded by agitators. Keith McIvor kept them at bay with the help of the operational staff, all of whom were hardened oilfield men. They encouraged the police, who were out again with the dogs. It was tense and stayed tense until the evacuations had settled down. But worse was to come.

Tribal integration, which had been going on in Nigeria for well over a century, began to unravel on a massive scale. The East was the primary destination for refugees who were fleeing their homes in fear of their lives. Frederick Forsyth's book *The Biafra Story* estimates the exodus from the North at one-and-a-half million, and the West at half a million. The impact of these returnees on their villages and relatives was colossal. Dreadful stories—supported by independent eyewitness accounts—of slaughter and destruction became more and more common.

After the small Hausa community had been shepherded onto a train under police escort and taken North, there was time for us to evaluate the situation and what might happen next. The idea of secession, its logic boosted by the terrifying experience of the pogrom, began to look like the only solution—despite the likely consequences.

*   *   *   *   *

While the East simmered, the movement towards reconciliation gained some momentum. Keeping Nigeria as one country had strong diplomatic support from the British and the Americans, and by the end of 1966, the scene was set for a conference in Ghana involving the key figures.

The country's future was now in the hands of the head of the army and Head of State Gowon and his four regional military governors, Colonel Adebayo from the West, Lieutenant Colonel Ojukwu from the East, Hassan Katsina from the North and Chief-of-Staff David Ejoor from the Midwest. The navy and the police were also represented, as was the territory capital, Lagos.

Agreements were made on repatriation of troops and outstanding salary payments, political structures and control of the army, refugees, their rehabilitation and compensation. A further conference to discuss constitutional affairs was pencilled in—but it took no time at all for the patched-up agreement to fall apart and the tensions to return. As 1966 ground on, divisions widened. The drive to secession was unstoppable, and it happened on 30 May 1967 with a declaration of independence made by Lieutenant Colonel Ojukwu. Less than seven years after Independence Day, all the hopes and dreams it represented lay shattered.

*   *   *   *   *

On the conclusion of our first collective contract with the new union, a brief triumph for our team, I moved to London to work on the Nigerian desk, which provided the link between the Group's executive director and the management in Nigeria. This gave me time to reflect on what had been a brutal political and societal revolution, and how it had diverted and finally halted our operation.

It should have been clear that the achievement of national independence, predicated on the assumption of successful regional independence, would reinforce tribal blocs and make nation-building a matter of alliances rather than a collection of individual voters. At the root was a flaw that subsequent governments would try to address through the creation of multiple states.

The national strike in 1964 had shown that the union movement could gather widespread support, but it had no political agenda and it was not like the Soviet revolution, driven by political activists. The few activists I had encountered in Port Harcourt had neither the determined ambition for change nor the ability to stir the people. In an environment of indirect rule there had been no spur to harden the revolutionary's resolve. No Lenin would come out of this bunch of leaders.

Our own preparations for disintegration had fallen short, and we had failed to develop an alternative export facility, which had damaged us financially. We had planned for everything we believed physically might happen, even the evacuation of the women and children from Port Harcourt. The families were escorted to the refinery jetty at Okrika to bid farewell to husbands and fathers. They spent a 'sunny day' on the creek heading for the Bonny export terminal. They then went by tanker to Fernando Poh, an island administered by the Portuguese, and joined ocean-going tankers to Spain, from where they flew, exhausted, to Heathrow. There they were met by their husbands, who had flown in comfort in VC10s direct from Port Harcourt. Our secret evacuation plan had been effective in its concept, but for the families, unhappy in its implementation.

*   *   *   *   *

It would be unfair to blame the company directors for any shortcomings in our performance in Nigeria. It had been a magnificent technical achievement by leaders who were at the peak of their profession, two of whom went on to head the Shell Group. Peter Baxendell and Lo van Wachem were exceptional men. Stan Gray and Dave Fleming were outstanding leaders, and in his period, Fleming was a political force. He was the only one of those extraordinary men who could or perhaps even would have stepped into the political maelstrom in 1967. During the national strike he had flown to Maiduguri to bring the prime minister back to confront and ultimately deal with it. But had the technical dream been a dream too far? My first career span in Nigeria tells the story of 30,000 barrels per day in 1960, 1,300,000 barrels per day in December 1966, zero barrels per day in 1967. It had been a whirlwind ride.

Looking back, even after 40 years or more I do not understand what motivated these majors to explode in an orgy of murder. Disgust with conspicuous consumption and widespread corruption might begin to explain one or two of the assassinations, but not all. There must have been something else: some outside agency perhaps that drove those young officers to act and then step back. Was there something in their background, the way they conducted their lives emotionally and spiritually? This was not an officer revolution, nor was it the spearhead of a movement involving other senior disenchanted officers. There seems something almost messianic about it. Someday perhaps someone will uncover the motivation.

Seen from far away in the Shell office in London, Nigeria was an increasing mess as the months passed. All the expatriate staff had left the East, Stan Gray being the last. He had been the guest of Ojukwu in an Enugu hotel for a few days. This was not as dangerous as the press suggested, particularly as Ojukwu's father, Sir Louis Odumegwu Ojukwu, had been one of our directors. The Biafrans were aware of Shell's position and our determination to remain impartial.

Some months later Michael Mann, formerly a vicar in Port Harcourt and then an Anglican bishop, approached me to ask if Shell would make it crystal clear that the company would not offer the Biafrans support, financial or otherwise. He put his flat at our disposal for the meeting with Christopher Mojekwu, a top man in the Biafran hierarchy.

The Shell camp was reluctant to have such a meeting, particularly against a backdrop of rumours about French involvement in oil concessions but, we reasoned, if an unequivocal statement of no involvement would help, then for humanitarian reasons we should do it. We set up a meeting under the aegis of shareholder, not operational business. I was to attend with Dick de Bruyne, one of our Dutch directors.

Dick was nervous about this, fearing the consequences of any leaks to the press. Everything stayed calm as we walked to the flat, my man continuously surveying every street, alley and closed door. Just four doors from the flat there were two black men in a trench in the pavement, one wearing an earpiece. This was almost enough to make us abort the mission. Dick eyed them with mounting suspicion. As calmly as possible I leaned down to them and asked, 'What's the score?'

'Two wickets down for 65,' came the reply, and to Dick's palpable relief I explained that these were Caribbeans listening to the cricket, not Feds eavesdropping at the Most Reverend's door.

The courtesies were brief. Mojekwu had great presence and made his points about the tragedy of war and his young nation's hopes, but admitted that they could go forward only with friends' help. This was where my man came good. He explained our position and our willingness to support any reconciliation. Beyond that we could not and would not take sides. Mojekwu asked, 'Is the door left open?' Michael looked at me, trusting that we would give no false hopes, no equivocation.

Dick said simply, 'The door is shut and will stay shut until there is a reconciliation.' There was no rancour, no argument, only a polite acceptance, and we left.

Shortly after this expedition I heard I was to be posted to Nairobi, so my seven-year involvement with our second home was coming to an end. But, like everything in Nigeria, there is always another chapter, another epilogue.

My secretary told me she had made an appointment for me to see Mr Ken B. Tsaro Wiwa. 'KB' had been one of my burdens in Port Harcourt. A garrulous, amusing, perceptive person, his pen was even sharper than his tongue. My association with him had begun in his student days. He came from Ogoni, which was a riverine province—the swamp of the Niger, where the dry land meets the wet. My eldest son and I used to fish in that area (without much success), and some of the local people would help us to get the boat into and out of the water. When Ken turned up in my office for his scholarship interview, he was surprised to find I had a little knowledge of his area. He won a scholarship with ease. Throughout his years in university he dropped in to see me from time to time, and in general terms I had cautioned him on the dangers of political grandstanding without support.

Right on time KB arrived, dressed in a beautiful new suit and filled with acid humour and teasing comment. When it was my turn, I complimented him on his suit and his obvious affluence, to which he replied, 'I have a new job,' and proceeded to tell me he had been appointed governor of Bonny as federal troops had successfully occupied the area. Brigadier Adekunle, the 'Black Scorpion', had supported his nomination. At that point I stopped the banter and the conversation took on a serious tone.

'Have you understood,' I asked, 'that you have thrown in your lot not with the Feds but with the soldiers? Soldiers know only yes or no, white or black. There are no greys, no nuances of language. "Shoot" and "Hold your fire" have an immediacy you're ill-prepared for.' I urged him to think again. He looked at me with his large eyes and, unusually for him, said nothing.

After finishing our coffee, we shook hands and he left. Until the next time, we said, which would come 15 years later. I had by then made my way to the East and kept in touch with Nigeria only through publications and friends. Ken's reputation had grown in the literary world but his fascination with politics had moved into dangerous territory. By the time we met again he was no longer acting as an individual but as a representative of a potentially disruptive political force.

*   *   *   *   *

The conduct of the war and the tales from it had worn both Jo and me down. Jo worried about her close friends and the colleagues she had been working with on her book, and about what would happen to her welfare initiatives.

Looking back on those seven years in Nigeria it is obvious that we were in the mainstream. In 1960 everything was on the move. Our new country was on its way to independence, its universities and schools were full. The newspapers published instant reports and commentaries criticising government (and there were already four regional bodies in operation), corporations and businesses; nobody escaped. The roads were so packed that minor accidents were frequent; you just moved on, shouting and gesticulating. When I waved my stump out of the window, the fuss would pause long enough for us to take advantage. Markets were brimming with local and imported goods, and price was set by negotiation, not by statute. Thieves and robbers plied their ageless trade just as they did in London. All in all, after the peace and order of Brunei, it was fun and full of life. Too much for some—but we were young!

Douglas was growing fast, physically and mentally. His arguments were tested in the Nigerian environment of debate and dispute to the enjoyment of all around him—he was more Nigerian than British! Our second son, Paul, had made his entrance in Port Harcourt and his nature was a happy one, which helped the mood of the household.

The political progress in those seven years was zero, but this went unnoticed by many of those working in the oilfields—there was too much oil to find and move to market. The scale of the challenge was all-absorbing, engaging the talents of our best and experienced engineers and technologists, sustaining morale and positive industrial relations. But among this cohort of expatriate adventurers there were those who had memories of Nazi activities and, more recently, Indonesian disruptions, and when the moments of real crisis came, they were sure-footed in their responses. They had seen too many lives taken to stand by and let it happen again.

4

East Africa and Back Again

East Africa was totally different from West Africa. Gone were the Niger Delta and the swamplands. Gone was the overwhelming density of people, the colour, the noise and the bustle. Here was Nairobi, a beautiful city with gardens, quiet streets and equable temperature, surrounded by open country in which antelope, gazelle, lion, cheetah and leopard roamed. It was a traveller's paradise.

The people were different too. Quieter, even reserved, they were polite and respectful. The brash mammy wagon with its bumper sticker yelling, 'Be lucky—let me have mine!' was nowhere to be seen. Here you could imagine journeying south to the Cape and north to Cairo, whereas in Lagos you might make Ghana to the north-west, the Sahara to the north and the dense jungle of Gabon to the south. In the West, you had geographical containment; in the East, you had freedom.

*　　*　　*　　*　　*

My job was personal assistant to the chairman. We looked after Kenya and its refinery, Uganda, Tanzania, Somalia, Zambia, Mauritius and the Seychelles, with Aden thrown in. We supervised their financial performance, agreed their budgets and helped grow their business.

But life was not as simple as that. ENI, the Italian government group led by Enrico Mattei, had set about rearranging the comfortable furniture by offering governments in these emerging economies equity deals and new refinery projects, so we needed to defend our investments. I was to coordinate performance appraisals and prepare for the restructuring of our interests. In an industry facing an ownership challenge in the post-colonial '60s and '70s, the challenge was not just to run downstream operations successfully, but also to re-establish revenue to secure profitable survival.

Mike Pragnell, who had spent his war in a POW camp after Dunkirk, was a superb mentor. He taught me the art of devilling, the in-depth inquiry that brought to the surface issues you never knew were there, and working with him honed my financial assessment skills. His team had been carefully chosen as a central directing group and had real depth. The Mombasa refinery deal was textbook tidy and Tanzania, with its 50:50 government/Shell equity share, was equally straightforward. The ENI initiative was losing steam.

Zambia was much more fun: Kenneth Kaunda's Mulungushi declaration on the nationalization of the means of production had raised the stakes. It was no longer 50:50 but 51:49—leaving Shell in a minority position. Our negotiating team consisted of a London director, John Francis, who had long experience of East Africa, a London lawyer, and me. Buoyed with the Tanzanian success, the lawyer in particular thought it would be a walk in the park. He was quickly put straight.

Andrew Sardanis, a Greek Cypriot who had just completed the huge copper-mine negotiations on 51:49 gave us the time of day—just about. Brushing aside the Tanzanian documentation, he said brusquely that he could do this in half an hour on the back of an envelope. Our leader protested but Sardanis made it clear that 51:49 was government policy and there was no flexibility. With that he handed over to his second-in-command Andrew Kashita, and left. This was not rude, but it was very close.

That evening John opted for bouillabaisse for dinner. I wondered why, in the middle of Africa, he would choose a dish packed with seafood, and unfortunately my scepticism was well founded. He took to his bed for three days. I returned to Nairobi, by chance sitting next to Kashita on the plane. After a few preliminaries we did the deal on the back of an envelope: book value, an energy supply agreement and a 51:49 equity share. I outlined the deal to London; it was done with minimum fuss and was still running well into the new millennium.

In Uganda, Obote had gone, Amin had taken over and his soldiers were on double pay. Africans were more at risk than the Europeans, and each month I withdrew a few more senior people. Nigerian soldiers were disciplined and respectful, but this was not the case in Uganda, and things steadily worsened. Amin's days were numbered and we waited. In the end he disappeared, leaving in tatters a country which, at independence, had been well set.

Further north in Somalia, things were deteriorating. It started with the government demanding the removal of the finance director, Bozzi. We protested, and our acting MD went to meet the prime minister to make our case. Using presentation skills built up over years of marketing in distant lands, drawing even on the papal knighthood that he had been granted during the Pope's visit to Kampala, he made the case. The prime minister apologised and said his people had got it wrong; it was not Bozzi who had to go, it was Pozzi. This was even more serious, since the luckless Pozzi ran the aviation operation. Our man was flabbergasted, but his ammunition was spent. Pozzi, thankfully, was replaced immediately; our acting MD left, exhausted, on a well-earned retirement.

No sooner had our new MD, Hugh van Drimmelen, got his feet under the desk than Somalia announced nationalization. When Hugh gave me the news, I asked him to check that the reams of political reports he had spent so long compiling had been destroyed. He said they were in the safe—however the key to the safe was with his wife, Anne, in Switzerland. The prospect of having Hugh plus his management team thrown in gaol was beginning to look likely, so we set about Mission Safe-break.

I waited for Anne—and hopefully the key—in Nairobi. From there we flew to Addis Ababa, but not before I had swapped the entire petty-cash float for a shelf-full of glossy magazines. We booked into the hotel, had dinner in an empty dining room and retired. I was woken at 7 a.m. by an unholy racket emanating from Anne's room: the bellboy had brought up two teas, which she had taken as an outrageous slur on the character of a happily married woman. Hastily I dressed, crossed the corridor to calm all parties and retreated—having secured the offending cup of tea for myself.

We were strapped in for the next leg and were taxiing for take-off when the plane suddenly halted at the end of the runway. The door opened and a Shell aviation superintendent boarded, strode over and handed me a large brown envelope. It contained a dossier marked 'How to Handle Nationalization', which was not only superfluous but offensive. My co-conspirator said she would dispose of it down the loo, but when this didn't work she returned looking pregnant with the words of non-wisdom safely secured in her knickers.

Hugh met us back at camp, where Anne set out the next phase of the operation. She would dress in mourning: severe black, stockings, hat and gloves; the Mercedes would fly the Dutch flag and she would proceed alone. We were to follow in a pick-up. I was not going to miss a minute of this.

Once there she sailed into the office, gave the civil servant in charge a cursory peck on the cheek and explained she had come to retrieve Dutch property from the safe. The inner office was opened. She requisitioned two boxes, filled them with the files and called for assistance to carry them to the car. The gleaming Mercedes then swept her and the boxes to the US Embassy, where all the files were destroyed. This was bravura at its best.

I had ordered a charter plane to fly the families out of Somalia, but our safe-cracker remained with her daughter and an Italian friend. Her cheetah-skin coat was not yet finished, she explained, and she would not leave until it was. And so women and children first became women and children last, and my boss agreed that it's better to accept the reality of life.

*   *   *   *   *

In Kenya there were still the remnants of 'white man' privilege. Until only recently, clubs and even some hotels had been for whites only. Society took itself very seriously and there was dignity on all sides. Civil servants wore suits, often with waistcoats. In the East, unlike in Nigeria, Africans, Asians and Europeans rarely mixed, and they socialised almost exclusively within their own groups.

In buttoned-up Nairobi it was tempting to be mischievous. Our first house was beside the Dutch ambassador's compound, which was enclosed by a high steel fence. At night a pack of Ridgebacks—large hunting dogs—patrolled. Their shift ended in frenzied uproar at 5.30 a.m. as the workforce began trekking to work, rattling the fence with sticks as they went. After we had been there a few weeks I was asked how I was enjoying the dogs. 'No problem,' I replied. I explained that I had dipped my stump deep into a jar of tomato sauce and run up to the house waving it, shouting to them to call off the dogs. After that the Ridgebacks were kept well away from the fence. It was typical of that society that the story went around the circuit like wildfire. Over 30 years later I was asked if I had really had my hand bitten off by a pack of dogs in Nairobi!

Our three-year stay in East Africa was sadly too short to forge the kind of bonds we had in the West—but not too short to gain another son, Robert Michael, and many friends. Nick Muriuki became chairman of Shell in Kenya and a leader of the commercial community. His great friend, Mwai

Kibaki, became president of Kenya, and Duncan Ndegwa the first African (and longest-serving) governor of the Central Bank of Kenya. Duncan had been at St Andrews with Jo and me. Philip Ndegwa, permanent secretary of the ministry of finance, was my handicap-four golf partner.

In the course of our business we met Tom Mboya, a Luo who was the natural heir apparent to Jomo Kenyatta. In 1969 Tom was shot and killed by a Kikuyu; the murder left in its wake deep distrust that only time—and it will be a long time—will wash away. So it is with the Scots and their neighbours: 600 years have muted the violence, but not the words spoken and unspoken, at times by people from whom you would expect better.

*   *   *   *   *

From afar it had been hard to follow what was happening in Nigeria, and the stories were at times too outrageous to believe. But when news came that we were to go back there, we jumped at the chance to play our part in the rehabilitation. I was to run the oil and chemical marketing company.

Arriving at Lagos airport felt like coming home. We were guided through the packed customs hall by a beaming Yoruba by the name of Ben who seemed to know everyone. It was good to be among the noise, the crush, the extravagant colours again; the weeping and sighing as arrivals were clutched to the bosoms of their welcoming families.

During our absence, Nigeria had overcome Biafra in a bloody extended war, eventually subduing a proud people who were by then crippled with famine and disease. With peace came reconciliation and, led by Jack Gowon, the country restarted as if recovering from a bad dream. Mobility across the tribal boundaries was improving and the oil industry in our old home of Port Harcourt was re-establishing itself. Oil and money would start to flow again, schools would re-open, markets would flourish. Hope and expectation were on the rise.

The next seven years in Nigeria were to be about rebuilding confidence and trust. A civil war fractures relationships and leaves behind a store of resentment but, with hard work, bridges can slowly be rebuilt. 'Speaking of the bitterness' is the first stage of rehabilitation, and listening is an important element in that. Here again Jo was a strong influence, listening to the travails of the wives and understanding and empathising with their concerns and fears.

My job was to fuel this economic recovery via 24-hour service stations. They were supplied by a fleet of coastal tankers that filled lorries and rail tanks around the clock. Each part of the chain across this huge country was vulnerable to equipment failure, miscommunication, operator error, theft and fraud. The task was gargantuan, and all the more so because the workforce was reluctant to stray far from home. Memories of the pogrom and war were still too fresh, and initially confidence was low.

*   *   *   *   *

The large colonial-style house in Lagos that was to be our new home was set in a well-stocked garden. Tom the cook was an ancient Bini whose first job had been with the railway project team as they made their way north 30 years earlier. He had cooked their meals in a converted kerosene tin. His wife, Madam Tom, was also from the Midwest. She presided over the servants' quarters, achieving some financial—if not necessarily culinary—success by dispensing beans to residents and passers-by from early in the morning. To this happy group was added Mary, a proud nursemaid who brought tone and style to the place. The running of the garden had been left to Okon, but on reviewing Okon's performance, Tom felt more was needed. When Moses, our man from Port Harcourt, appeared on the doorstep he was hired immediately as major domo. Hyacinth followed, and the team was completed by Ezekiel, a second gardener who appeared as if by magic. Another ancient with wizened features, he promised horticultural expertise that placed him somewhere between Charles Darwin and Percy Thrower.

Ezekiel's philosophical pronouncements became legendary in our household. At one point he was keen to have the address of the United Nations' secretary general, because he, the secretary general, needed to be informed that the gardener of his representative in Lagos had gone jig-a-jig with Ezekiel's new and much younger wife. On Paul's departure to school in the UK, Ezekiel mourned: 'A tear fell from my eye as the compound grew cold.' When some disobliging insect ate through his wilting cucumber plant, his rather less stoical comment was, 'The damn fucking thing done disappoint me!'

The houses were well situated but fairly old. One day we had a domestic problem with a lavatory which would not clear. The small contracts division arrived, disconnected the pipe to the septic tank, cleaned out and replaced all moving parts and reconnected us. The engineer called for a test. My elder son obliged, the toilet was flushed … but there it was, still floating undisturbed in a full pan of water. For a moment the engineer paused, then with great seriousness turned to my wife and opined that 'the faeces are too light'. Liver on the menu would have to solve the plumbing problem.

A household in Africa gives an insight to the life of the men and women in the street, their sadnesses, their happinesses, their events, their lookings forward and their lookings back. Death is always around the corner, sickness is not far away, and a day undisturbed by either is a day for celebration. This is what provides joy in a sea of misery: Africans do not sink, they seem automatically to rise, and their buoyancy seems infectious.

But life in Nigeria left no time for lengthy contemplation. The military government wanted the country on the move. The first priority was to unblock the port of Lagos, and Brigadier Adekunle was allocated the task. He began with the traffic jams that clogged the way to the docks. There was no discipline on the roads; lorries approached roundabouts four abreast and none gave way. But a couple of rounds from the brigadier's machine pistol spread the message quite efficiently and the heaving mass of diesel-powered mammy wagons suddenly became orderly lines. The union problem was solved by throwing the six-man executive into the harbour. Paperwork got similarly short shrift—nothing had to stand in the way of clearing the docks. Customs were swept aside as goods evacuated warehouses at speed. Adekunle was a logistics genius—but a financial disaster. When he eventually heard about customs, he realised what an opportunity he had missed.

Gowon, a Christian from a minority tribe, personally carried the reconciliation movement forward, welcoming returning refugees and guaranteeing their safety. This set the tone and the vast, complicated country slowly began to pull together again.

*   *   *   *   *

My office overlooked the harbour and the milling throng at the bus station. Further along the street were the offices of Shell BP, my previous employers, who were slowly getting under way again after the suspension of their activities during the civil war.

Our cash flow issues lay not in sales, but rather in the difficult province of collections. Customers took refuge in the confusion—claiming absence of invoices, disputing deliveries or falling back on the ever-popular 'the cheque is in the post'. One day I went downstairs four floors to see who was in charge of invoice reconciliations. I was concerned about some contractors—reputable international companies—that had very large outstandings. As I waited in front of the manager's desk, I practised the patience I had learned from my ten years in Africa. His telephone conversation was interesting: he was obviously importing goods and distributing them across the country. It sounded like an active business, and he was a hard negotiator, driven, it seemed, by his financial exposure. After ten minutes he rang off, turned to me and asked gruffly, 'What do you want?'

'To meet you,' I said, 'and to understand what you do or do not do here,' and with that I introduced myself. I went back upstairs and reported to the personnel manager, Ajose. After a swift investigation and a reluctant admission, our rogue importer was on his way to conduct his shipping business elsewhere.

That same day I met by chance Elizabeth Uzuodike, an Ibo whose father was a respected bishop. Elizabeth was a qualified accountant, a quiet, articulate, wise woman who at the time was unemployed. I invited her to come in and meet Ajose, and by the end of that day she was the new head of customer accounts. In two weeks she had, by careful analysis and reconciliation, discovered that the international companies' payments had been credited to a group of local contractors who by now had almost a year's credit. It would appear they had in return been funding our ex-employee's distribution business.

In my East Africa role I had overseen a diverse group of marketing companies and run a supply system from a refinery with a wide distribution network. It stretched through Kenya and Uganda, on to Rwanda, Urindi and (sea-borne) to Mauritius, Somalia and Tanzania. Assessing the profitability and return on capital of these operations was complicated by the differing tax regimes and currencies. Working with individual management teams, we identified weaknesses and shortfalls in performance, and then expected the teams to fix them. I now sought to put this exercise into practice in my own ship.

First I wanted to see the operation on the ground, so I travelled far and wide visiting service stations and depots to inspect the quality of the staff, the safety of the product-handling, the maintenance of valves, defence against dry-season fires, cleanliness of the toilets, and security of cash systems and gas-cylinder storage.

Stock was another important area, as my Hausa Northern manager discovered after our conversation on his appointment. Reconciling the stocks with the records was essential. As he travelled north into his territory he stopped at a depot to discover that stock could not be matched with records—because there was no stock! He rang me to say, 'I have the men in gaol. No stock, but great confidence in the future as I am for the time being the depot manager as well as being the new Northern manager.'

The main installation at Apapa in Lagos was the heart of the operation, and the pressure was on for its swift economic development. As the federal government had initiated a huge road-building programme, our bitumen facilities were on 24-hour operation. As always with heated product, safety procedures were demanding and had to be inspected regularly. The lube-oil blending plant was a technical operation and its quality control was also vital.

I never had any difficulty seeking help where and when it was needed, but it had to be the right person who would identify the problem and then work with the management to find a solution, not just add to the pile of reports gathering dust on the installation manager's desk. My man John Hughes did just that. He wrote nothing but a list of action points, and regularly reviewed progress. In short order he created the confidence and dynamism that generated and sustained continuous improvement.

From my first spell in Nigeria I had learned to challenge, to accept nothing until you had proof. Some people were suspicious of this, fearing it would lead to intrusion or criticism. Pondering this as I looked out the window at the bus station below, I heard the cry 'Stop! Thief!' I could see a stream of travellers running and pushing out of the station chasing a nimble youth. He easily outran them. Disappointed, the pursuers made their way back to the station and their luggage—which by then had been ransacked by the nimble youth's nimble-fingered accomplices. The sting had happened before and would again to travellers the world over.

The arrival of a hard-working young IT manager interrupted my reverie. He wanted to talk about taking on a wider role. I listened carefully and then explained that, unlike his impressively tidy and neat operation, the world out there did not run with mathematical precision. Our world was more like the bus station, where you had to be constantly on the alert. I used the example of Kano airport, which was at that moment dealing with the Haj, the Muslim festival that prompts thousands of Nigerians to fly to Jeddah. Our tanks were emptied almost as soon as the product had settled, and a stream of tankers made their way continuously from Lagos and back. Flights to and from Jeddah were round the clock, so quality control was on 24-hour alert. This was very good business provided you collected the money. So I suggested that the young man should go and take a look, and by doing so bring alive some of the figures he processed. My last words to him were, 'Be inquisitive, and don't be surprised.'

Two days later he rang from Kano. 'They have not passed through the last price increase! I found the notification still in the in tray with no action taken.' He immediately set to and put it right, and after that experience the figures he saw on his screen meant a lot more to him and his explorations met with more success. He had become an active team member rather than being just a technical operator.

Distribution and sales organizations need to know their customers as well as their market. One of the chemical organization products that sold well was Shelltox, a mosquito-killing spray. Mosquitoes could infest your house if you did not keep the doors closed, and though they were not malaria-bearing, they stung like angry wasps. So every house had its red bottle of Shelltox and its pump spray or, for the more sophisticated customer, an aerosol spray. The sales revenues for this product never faltered, they just kept moving upwards ... until without warning this happy state of affairs changed and sales began to fall away. Marketing efforts were redoubled with little effect.

Then one day at the local supermarket, I overheard a Nigerian inquiring where the Shelltox was. The sales assistant gestured to a shelf bearing an array of blue aerosols. 'No,' said the customer, 'I mean the red ones, the mosquito-killers.' There it was. The organization had adopted the new Shell Chemicals colour scheme of powder blue and the market reacted: no blood red = no kill. Immediate reinstatement of the red cans led to an immediate reinstatement of the steady sales growth. Better to listen to a customer than blindly follow 'Group Think'.

Each year we sponsored the Shell Belt, a trophy awarded to the best amateur boxer in Nigeria. Originally suggested as a way to channel construction teams' energy into a programme of sport, the competition had been created with the guidance of Olu Oguntokun, our bitumen manager and an Olympic referee. As a corporate event it attracted much public attention and opened many doors. My father, winner of several amateur boxing cups, would have approved.

In Nigeria, boxing is a well-organized professional sport and the country boasted two world champions, Hogan Bassey (featherweight) who was to judge our tournament, and Dick Tiger (middle- and light-heavyweight). Our bouts started at flyweight and progressed through to heavyweight. The action was tightly refereed and any cut was immediately examined by the officiating doctor, Innes Palmer. That night he stopped one fight. The winner of the light-heavyweight bout, Isaac Ikhouria, who went on to win bronze at the 1972 Olympics and gold at the 1973 All-Africa games, won the championship that evening. It was my job to present the prizes and the winner's belt. The first part was easy but the second, fastening a heavy belt around a sweat-soaked athlete's waist, was not. I was thinking I would have to use my teeth when Olu came to my rescue.

The next day when I thanked Olu again, I asked him about the boxers' prizes: for the winner a hand towel, and for the runners-up a biro. I expressed amazement that these guys would battle away at each other to receive such unremarkable prizes. Olu explained, 'It's the prestige, not the prize. They got their invitation, that's enough.' I said we should do a better next year and give them something they could put on a shelf, and we did.

*   *   *   *   *

Selling pesticides to cocoa farmers through rich intractable mammies could be a colourful commercial experience. When the Group's director for chemicals visited, we arranged for him to meet one of our biggest customers. She was ready for him. At 10.30 a.m. on a sweltering morning the austere Dutchman introduced himself. From inside her Kokotine shop the woman produced four of the finest crystal glasses and a magnum of chilled Pol Roger. She opened it with a flourish and filled the Dutchman's glass. Gesturing at a man who was skulking nearby, she said, 'He's the press—he'll get nothing until we see what he writes.' This totally threw our visitor who mistrusted all reporters, never drank before 7 p.m. and was already rather taken aback by the woman's considerable physical presence. After his first nervous sip, he made a tentative stab at conversation.

'Have you family?' he thought would be pretty safe.

'Yes, I have a son at Harvard finishing his MBA, a daughter doing medicine at Oxford and another at Stanford.' Further confounded, our man knocked back a second glass, at which point I felt it was time to get him out of there.

As we pulled out, the intrepid reporter tried his luck: 'A statement, please!' The Dutchman blanked: faced with the representative of this Ibadan daily that had a circulation of over 1,000 readers, he was struck dumb. We intervened, gave him our visitor's name and a picture, leaped into the car and sped away, leaving him to try his luck with the remains of the magnum.

*   *   *   *   *

Running a network of petrol stations is all about attention to detail. The dealer needs to be dedicated to his outlet and focused on his business interests. Any time the pumps are not in use is wasted time. Using the station to sell other goods dilutes its effectiveness and clogs the pumps—and while that's going on, potential customers drive by.

Lubricants were a real sales opportunity because cars in Nigeria were older than average. We ran an advertising campaign around a mythical bespectacled, white-coated, expatriate technologist Dr Sabi Sabi—the man who could 'doctor' your car and keep it fit. This did well and we followed it with the Golden Chip promotion—if you found a golden chip inside your lubricant bottle, it would be worth 100 gallons of free petrol. I went over the plan with the team. It was essential that we had stock in place, because as soon as we advertised the chip there would be a run on the product. The installation plant had to gear up to produce and fill the bottles. We went through every step in the chain. I said we did not want all the chips ending up in one street or village; a trusted member of staff had to ensure countrywide distribution. In the meantime, until everything was ready, the chips should be locked in the safe.

The big day arrived and Dr Sabi Sabi was all over the papers extolling the benefits of the lubricant and the reward for common-sense buying. Sales ballooned and we were delighted. After two months I called for a breakdown of the winners, but it was slow in coming. When I called again, a shamefaced lubricants manager admitted that security had been so tight the gold chips had never left the safes. There had been no dastardly diversion of the distribution—there had simply been no distribution!

We made sure Stage Two of the campaign highlighted the winners, and this gave it the final boost, fulfilling all our hopes. This time every chip had gone—and countrywide.

By the early 1970s the military government began a road- and airport-building programme that made us the biggest bitumen supplier in the Group. This truly stretched us; we virtually lived in the bitumen plants. There were bottlenecks every day, and we chased rail tanks and wagons through the network, pursued the return of our road vehicles from the customers and charged heavy demurrage, which added to our already attractive margins. The business was a bonanza and our facilities and teamwork gave us a head start.

We had a heavy share of the army business and an even heavier share of aviation's. Payment was a problem, though; if you were going to get anywhere, you had to have proof of supply. In the end I visited the permanent secretary of the ministry of defence, a Northerner and a most effective administrator. The interview was tense and my line was not a threatening one, but one of setting the standard for commercial transactions and safeguarding the country's reputation. This found its mark. Though I was summarily dismissed, the cheque clearing the arrears arrived the following day.

The ministry that looked after us was run by Meshach Feyide, who subsequently became OPEC secretary general. When I met him years later in Bangkok, he described to me how he had escaped being taken hostage when terrorists attacked the committee and took country representatives for ransom. He said he had seen a movie where a hunted man stood on top of the lavatory so that the terrorists looking under the stall doors would not know he was there. Meshach followed suit and the men left without him, so he was free to coordinate subsequent events. It had won him a reputation as the 'Lavatory Hero'.

Meshach was a man long in advice but carefully short in action. Bearing in mind the brittleness of a military government and its propensity to act first and ask questions later, this was wise. There was, however, one issue that would not go away: price equalization. The government wanted the same price displayed throughout the country. My predecessor had tried unsuccessfully to dissuade them with all the economic arguments, but quoting Adam Smith had got us nowhere. I decided to take a different tack and suggested we design the scheme before it was imposed upon us. The structure for price equalization is a simple one—sales made further away from the point of distribution receive subsidy, and those nearest pay more. The subsidies are paid into and out of a fund, which is administered centrally. The important point was that we should pay 'on the net', that is, subtract our payout from our pay-in. The worst outcome would be to pay into the fund and then have to wait for the payouts. Our experience of administrative weakness, delays and intervention by auditors made payment on the net a primary objective.

I called the industry together and explained the scheme and the political imperative, and we agreed the price equalization scheme I would present to Meshach the following day. He took one look at it and said it was not enough for a viable fund in an incompetent world, so he introduced an over-recovery element to protect the fund. But he did agree to paying on the net. And so with a great flourish the scheme was launched. With our large market share the money piled up, so for the time being we were cash rich. I pressed Meshach for an account to pay into. He demurred. My overdrafts were gone, cash had settled my supply debts and interest was now an addition to my bottom line. The Group treasurer sent me a snooty note asking me why I had cancelled the bank overdraft, followed shortly by another note inquiring why I had repaid Standard Chartered and questioning our financial management. The coordinator who reported our business to the directors asked if we were getting our supplies elsewhere as we now owed nothing—a most unusual state of affairs, and I had still no word from Meshach. At last he took me into his confidence. He did not want the fund to be available for political purposes, which would be a danger for the next few months. Until then I would run the tidiest balance sheet in the Shell Group.

The scheme was not without its dangers. The sales of our dealer in Maiduguri on the border with Chad had increased exponentially. Strong marketing and efficient delivery seemed an unlikely explanation, so I paid a visit and found that the station was just yards from Chad, where prices were not equalized. Drivers were filling up in Maiduguri and returning to Chad. When it got to lorries' as opposed to private cars' tanks, it was smuggling on a large scale—or, put another way, it was a beneficial extension of the price equalization scheme to a neighbouring friendly country. Like much else, where there is a river of profit it can be diverted, but never stopped. Such is the human condition.

Fortunately for the scheme, opportunities around the borders stretched our resources to breaking-point and filled our coffers, but it was clear we were not alone in the struggle to satisfy. Cement had also become a priority; ships carrying cement queued offshore awaiting their turn to unload, an image that drew international commentary. The government agreed to pay demurrage, which only lengthened the queues. Rumour had it that some owners parked ancient empty boats just for the demurrage—although that seems unlikely. Taking the children to see the fleet outside the harbour became a Sunday entertainment.

Our success did not escape the eye of the Inland Revenue, and I was summoned to my first interview with its chief. Looking at the size of my tax bill and the even greater size of my associate company's, he expressed sympathy and asked rhetorically,

'What can we do about it?'

'Pay it,' was my answer. 'If we fail to pay we deny citizens the hospitals and schools that they need. Otherwise how can we look our Maker in the face and say we have done our Christian duty?'

'Amen,' he sighed, philosophically, but that was not the end of it. Since we showed no interest in coming to a private arrangement, they slapped us with a turnover tax—for which there was no justifiable basis in law—the very next day. The battle was on.

At this stage we made an extremely good decision. We engaged Chief Rotimi Williams QC, who subsequently became the first Nigerian to become a senior advocate of Nigeria. Weighing in at well over 20 stone, he had sharp eyes and terrier-like tenacity. He said he wanted to see every piece of correspondence with the Inland Revenue the moment it arrived. He took it all, and each form was completed, returned and registered on the day it arrived. For three months, the ping-pong of manuscripts continued until time ran out for the Revenue; it was to the courts or not. Rotimi's impeccable management and reputation was too much of a challenge and he pressed successfully for the case to be dropped.

While this was going on, one of my accountants, under pressure from a junior member of the Revenue and in the interests of good relationships, gave the man a tin of tree-killer, a special product for farmers. The tree duly came crashing down … on top of the man's henhouse. In a way it was symbolic that Rotimi was bringing the chickens home to roost as the tree fell on their home.

*   *   *   *   *

Golf was a favourite sport for the military, and generals and air marshals made up our competitive fields. The browns—that is the oiled sand that made our 'greens'—were as testing as grass. The Nigerian Open became part of the African tour and attracted top players—Jack Newton won it one year in a play-off with a young assistant pro. As amateurs you could qualify for one of ten places, or one of four places if you could make the final two rounds. I made it through most years and one year I stood looking at my 76 on the board when a lady sympathised with me for my poor day; little did she know that was well beyond my expectation. As a place for breaking barriers, the golf club had little to match it.

Before independence, the country had driven on the left with imported British right-hand-drive cars, but as part of throwing off its colonial past it decided to shift to driving on the right. We were at a cocktail party at the racecourse just before changeover day. As we drank Champagne, cars drove round the course on the left and then, to celebrate the move, returned on the right. On the roads terrible crashes were forecast, and tangles that would tie up whole cities. One local newspaper launched the idea that lorries should make the first step in a two-phase operation, arguing logically that they were always on the right side of the road anyway. When the day came, policemen, Scouts, Guides, social workers et al manned every intersection … but that day and seemingly for several weeks afterwards, cautious motorists did not venture out, so the much-predicted doomsday never happened. All went unexpectedly well.

Not long after this we had a cholera scare. Inoculation booths were set up and households were required to present themselves for vaccination. It was at one such booth that our friend Judith, a highly qualified nursing sister who was respected by all, particularly the police, stopped our gardener from taking his third shot. He was already well on the way to a self-induced bout of cholera. We had him groaning on his bed for days.

Judith was called on at all hours to deal with all manner of emergencies. On one Saturday afternoon she was asked to come to one of the islands where a young lady was apparently in danger of being crushed. Judith found her wedged beneath the lifeless body of an elderly cook. It transpired that the young lady had not been paid for services rendered, and refused to move or to release the body until she had. Judith, not a slight woman, manoeuvred the gentleman, whom she recognized, onto a wheelbarrow. She opened her purse and paid the young lady, then proceeded to wheel the late cook back to his employer's house, where she advised on funeral arrangements, received reimbursement, and left.

Judith's gas boiler exploded one Sunday evening. We quickly quashed the rumour that the explosion was Biafran armaments: it was in fact due to corrosion and leakage. This was not the only unfortunate aspect. Judith's neighbour was none other than Meshach Feyide who, being an engineer, had rushed over in his shorts to help. He had been rudely ushered out by our expatriate engineer with the words, 'No garden boys here, please.' This was deeply offensive, and it was only thanks to Meshach's equable nature that good relations were not badly damaged. But it rankled.

When, in the early 1970s oil hysteria broke out, Meshach was at the centre of an oil auction. Bidders arrived from everywhere. Shell was not involved since we had our own production, but executives poured in from around the world with sealed and unsealed bids to try their luck. Deadline day coincided with a huge funeral procession for a 'second burial'. These elaborate celebrations, characterized by enthusiastic partying, were designed to ensure the deceased had a peaceful passage to the afterlife.

The streets were full of people, dressed in all the colours of the rainbow, making their way to the church, and Meshach was among them. As the submissions deadline grew nearer, anxious bidders sought him out and pushed through the crowd to press new and revised bids into his hands. Less-well-informed hopefuls pressed bids into the hands of other lean, distinguished-looking celebrants, much to their bewilderment. In the end the Japanese won with a bid, astonishing at the time, of $42 per barrel. In fact they never took delivery—common sense prevailed at last, but the funeral procession had certainly added colour to a spectacular milestone in oil pricing.

*   *   *   *   *

The political structure of the country we left in 1966 had exploded out of existence. General Johnson Ironsi had come and gone, spreading confusion and suspicion with his good intentions. Gowon, a compromise candidate, was a justifiable replacement. A competent staff officer, he had a strong grasp of process which he imposed on the mêlée below him. He created individual baronies which had power to administer their own local affairs, and these in time emerged as individual states. Competition between the States—which of them had new investment, which had growth, which had working facilities and courts of law and which dealt most effectively with its scandals—was a positive force.

Commander Diete-Spiff of the navy, military governor of Rivers State, built his own naval college along the lines of Dartmouth Naval College in England, which he had attended. Port Harcourt, thanks to oil money, recovered quickly. Mobolaji Johnson fitted the bill for Lagos, and the city was soon back on track. Growth—or at least car ownership—was so prolific that the streets were gridlocked. Some bright spark suggested a solution that would halve the number of cars on the road on any one day: enforce the use of alternate-day number plates. The idea was that owners could drive only on every other day, when it was their number plate's turn on the road. The plan fundamentally misjudged the ingenuity of the Lagosian, however, and soon there was a thriving market for alternative number plates. Congestion quickly returned to the extent that if you needed to catch a plane, whatever time it of day it took off you left for the airport in the middle of the night!

Brigadier Obasanjo, who subsequently became president, was in charge of supplies, and he facilitated our payment from the ministry. Obasanjo was like Gowon, except that he was not arrogant as some of his colleagues were, and he was grateful for help. A pragmatist and a team player, he took on the challenge of rehabilitation. Alongside him on the military side were two vastly different characters from places more than 1,000 miles apart.

John Obada, head of the air force, was a man large in stature, voice and personality. I first met him on the golf course. When we were on the 12th facing a tricky putt, a motorcycle roared up. Obada stood back, took the salute and asked, 'Are we at war?'

'N-no!' stuttered the despatch rider.

'Then return to barracks.' Obada sank his putt.

At the end of the game he announced the Kaduna Open, or rather the Northern Championship. On this course there were no walls and no ditches, but white lines demarcated the out-of-bounds line. To ensure fair play, John had mobilized his men. They guarded the line so effectively that when my partner on the fourth knocked it over the line and went to pick it up, he was shown the bayonet and sent on his way. The air force certainly had balls!

The head of the navy, Admiral Wey, invited me to lunch. After a tour of the facilities we sat down to the navy drink: gin with a little tonic. We talked about his home town of Calabar. He had heard of Eddie Duke's sad death and the saga of his burial, and expressed appreciation for the kindness and respect that had been shown. I talked fondly of Eddie and Thelma, not as employees but as friends who were part of our community.

Brigadier Godwin Ally was an active man and a loyal one. A ferociously competitive golfer, he would plunge into the bush with as much determination as a foot soldier in pursuit of an enemy. I wanted to make him captain of the golf club but David Ejoor, later a major-general and military governor, advised against it, saying, 'He doesn't have the competence.' I found this strange, particularly as he was acting governor of Lagos in Johnson's absence, but David explained, 'Some positions are easy when you're not too close to them and you have assistance; others are less easy, and you don't gamble with those.' This was truly Delphic, but deep down I think David did not want to put at risk a game that was so important to him. Fascination with golf spread everywhere, even in the most unlikely places. Commander Diete-Spiff himself was down to a four handicap, matching David's.

The armed forces took reconciliation and rehabilitation full on, and called a national conference to which I was invited as a businessman and investor. After addresses by the head of state and chairman of the bank, among others, we separated into syndicates. Brigadier Obasanjo led mine and appointed me coordinator and presenter. Refurbishment and development of the infrastructure were our top priorities: road networks, airport modernisation, and—crucially—investment in education. We needed to overhaul university and college provision. Ours was a fresh and exciting presentation, clearly appreciated by the forum.

Things were calmer and more peaceful under Gowon, and his reconciliation message encouraged progress. He and his team knew their weaknesses, and consequently recognized the importance of the civil service. But the administration was uncompromising. A radio broadcast invited spectators to attend a public execution on the beach at Lagos, where armed robbers were tied to stakes in the sand and dispatched by firing squad.

Corruption was still a big issue. Entrepreneurs rather than soldiers or senior civil servants exploited opportunities that arose as part of Nigerianization. Development threw up opportunities for contracting companies which they sometimes took advantage of on their own initiative, or were sometimes forced to. Our protection lay in the competitive market; we operated on the basis of optimum price and terms. Our business was also too public to meddle with: a strict process governed the award of transport and engineering contracts. But despite all this, some people became wealthy not from their activity but from their position. In this burst of chaotic economic growth we could know with confidence only what we saw—and as far as oil money was concerned, there was much we did not see!

Power corrupts and absolute power corrupts absolutely. Some corruption was exposed and addressed in this vibrant society, but the scourge never went away. As ever, the state of things is far easier to understand in retrospect when perspective is less distorted by the battle of simply getting through the day. This was a country in a period of turbulent growth, of great wealth, opportunity and political confusion.

As new wealth filtered through to the starved universities, it also sponsored a cultural confidence. The Black Arts Festival sprang to colourful life, rich in music and dance—and again in opportunity. Twenty Mercedes imported for the festival into the Port of Warri were to be driven by contract drivers to Lagos. Strange to say they never arrived, disappearing straight into the maw that is the Nigerian market.

An outpouring of new art began in Ibadan. Susanne Wenger, a Swiss–Austrian artist who had originally studied at the Academy of Fine Arts Vienna, was involved in setting up an artist cooperative in Osogbo. This produced many well-known artists, tapping a rich vein of talent that had quietly grown over the centuries from the days of the bronze-workers in Benin and the woodworkers in the Southern States. Products from the potteries of the mid-belt soon became much sought after. Creative and dramatic in its use of colour, the art was insightful and often unflinching in its choice of subject.

*   *   *   *   *

The price equalization scheme was working well. It had had the intended countrywide impact and our contribution to its design and operation was appreciated. The government's next objective was to take an active role in the growing economy. Participation in the downstream oil business would make a good start, so its first target was to take a majority stake in the refinery. This was a straightforward negotiation and completed without difficulty.

The government then needed take a position in the marketplace to demonstrate its involvement to the public. As market leader with a countrywide representation, we were the obvious choice of partner. In public eyes it was a good association. From our point of view, we wanted to defend and advance our commercial interests in what was a rapidly growing economy, so our objective was to get the best deal for the shareholder.

We, the marketing company, were approached to see if a joint venture deal would be possible, but this was not a clearly defined operation like the refinery, the market being complex and susceptible to external influences such as fluctuating demand—the needs of the bitumen market, for example. The company's operations were run mostly by Nigerian staff, with expatriate participation only in some senior management levels, so the prospect of government involvement was not widely welcomed. We counselled key staff to ensure their concerns were taken into account—in fact the situation had distinct similarities with the way nationalization was approached in the UK.

As a first step, the government asked us to evaluate our assets. Respected agents Knight Frank & Rutley carried out a full assessment, whereupon the ministry asked for a Nigerian assessor, from whom they expected a political price, not a commercial one. Hope Harriman, a well-known Midwesterner whose son played rugby for England, stepped in.

Harriman criticised the Knight Frank report, saying that it seriously undervalued our assets, and he was right. He was also adamant that residential property need not be part of what was a political sale. The whole purpose of the valuation was to set the price at which government would enter the new venture. Within 12 months, we sold our houses at open-market prices. Harriman's overall assessment embarrassed the government while apparently enhancing Knight Frank Rutley's reputation. The whole issue of book value was where it stayed while we prepared for a new identity and a new name. Royal Dutch Shell Chairman Frank McFadzean was uneasy when I said I wanted to stick to our position and the Harriman valuation for a 50:50 joint venture, and advised me to take care.

The logo for the new 50:50 joint venture, which would own and run the business, had to reflect the personality of the new shareholder: the government and Shell working together and jointly owning the asset. We created the green eagle motif inspired by the Nigerian football-team logo, and were delighted to avoid the image-makers' exorbitant bills. It hit the spot and had total public support. I passed the deal on to my successor, confident that we had put a new arrangement in place without disturbing the business's momentum or staff.

By this time the government was well advanced on the Nigerianization of foreign equity in banks, retail and construction. Barclays was upfront and put a proportion of its equity on the Stock Exchange. Jo and I advised our household to buy, and advanced them the means to acquire shares at book value in a profitable bank. When the time came they joined the cast of thousands at the Federal Palace Hotel for their first annual general meeting. I drove the investors down, Tom and Madam Tom, Ezekiel and Mary.

The meeting began with Gordon Thomson, a long-time west-coaster and then-MD of Barclays in Nigeria, introducing the report. After polite applause he asked for questions, three of which among many were memorable. The first questioner asked if there was a mistake on page 168: the figures did not add up. Gordon looked behind him to a towering bank of advisers who, having hastily riffled through the pages, agreed with the questioner. Great applause. The second referred to the report again, to Gordon's dismay, drawing attention to Barclays' lending to the agricultural sector and asking if the board understood that farmers never paid their debts. Stamping of feet. Gordon responded that he was confident that they would. Silence. The third referred to Barclays' record on Nigerianization and the number of Nigerians who were now managers. Didn't the chairman know that Nigerians were not to be trusted with money and that since it was now the shareholders' personal money that was involved, this was a very important consideration? Gordon made a stout and good defence of his Nigerian staff, illustrated with a number of excellent examples. Muted applause.

The Stock Exchange got under way, and with it came further opportunity. What if the holder of the equity were to be offered a much higher price, would he or she sell? Why not, if it satisfied the national criteria? In such a case it would surely be simple for the owner to guarantee a bank loan … and so, surprisingly, it was. Naturally, once one such operation has been successful, another follows. And so a multi-millionaire class was born.

*   *   *   *   *

During our time in Nigeria we had moved from colonial times to independence to secession and civil war and back again. We had lived with people who had lost everything and started again—as they had in Europe at end of the Second World War. Departure was not easy—how could it be when the permanent secretary of internal affairs had declared me a 'prohibited emigrant'? The chief of staff of the armed forces invited me for a last game of golf, and we said farewell to our friends around the country.

Events to mark the end of my tenure included a boxing tournament at the Shell club, and a motorcar rally. At the finish the cars, having chased 100 miles plus around Western Nigeria, were held three miles out of town while the boxing tournament came to an end. They then arrived at full tilt, all crossing the line within in a few compressed minutes of each other. Everyone seemed to win something.

I loved this frenetic activity. The chaos of unplanned action and the surprising outcomes made it such fun. It was 13 years of real life during historic times that I would not have missed; Jo and I were extremely grateful to have had the chance to go back.

5

Thailand

The Far East was not new to us, nor was the hustle and bustle of a tropical city where steam rose from baking-hot roofs when rain fell. But the gentleness and quiet manners of the Thai people were very different from the aggressiveness, riotous laughter and high-decibel life that we were accustomed to.

Until now, my only contact with the culture had been through the film *The King and I*. When I researched this further, I discovered King Mongkut and a Thai story that fitted well with my experience. The king, Rama IV, was the eldest royal son of King Rama II. He was born in 1804 during the first reign of the Chakri dynasty, which had founded Bangkok on the banks of the Chao Phraya river. He was 20 when his father died, and had just become a monk, so his brother took the throne. He remained in the priesthood for 27 years until his brother died, and during that time studied English, which opened the way to a knowledge of world affairs and science. Able to travel the countryside as no previous Thai king had, he became attuned to matters that concerned his future citizens, and his religious experience imbued in him an understanding of the ceremony that binds a nation. This apprenticeship made Mongkut a wise and popular monarch when he eventually ascended the throne.

The story of *The King and I* concerns the toleration of missionaries, and here Mongkut was far more enlightened than governments in neighbouring countries in the nineteenth century. In its way, this enlightenment seems to inform the Thai character, there is a tolerance and an ease—which, however, can have its limits. With the Nigerian temperament you can push forever; the temperature may rise, but it seldom boils over. Conversely I found with Thais that the temperature never rises *until* it boils over. To some extent my ability to gauge this forbearance determined my success in Thailand. I had advisers, and I listened to them.

Taking over new companies is about people first, and then performance. The marketing company had been in existence for over 80 years, having received the first shipment of kerosene that came through the Suez Canal in 1892. The tanker, *Murex*, was named after a carnivorous sea snail. It had an excellent network of service stations throughout Thailand, managed by experienced dealers who understood the business. The distribution network was good and the coastal fleet well managed with support from Shell tankers, but the terminal looked light on investment and the tank installation needed refurbishment.

The refinery was a partnership with KY Chow, and this was the fun part. KY had been a professor of wood technology before the Second World War. He had left China for the UK and helped develop the Mosquito bomber. After a successful war he returned to China, but in the increasingly disturbed political situation he decided to move again. He first tried Taiwan, and then Singapore, but found nothing in either to attract or retain him, so it was Bangkok.

He agreed to build a sugar refinery for the government and put it into profitable operation. Mission accomplished, he then performed the same industrial magic with a sisal plant. In the interim he traded very successfully on the international wheat market. Thailand's military government, having committed to building a petroleum refinery, were looking for someone to make it happen, so KY Chow of sugar and sisal fame was the obvious choice. He needed a partner to provide technological support and put in some equity; Shell ticked those boxes so the courtship began. McFadzean, the Group chairman who had spent significant time in the Far East, was supportive and a deal was done.

When KY came for the final discussion in London, he was presented with an agreement running to hundreds of pages. He leafed through it and said it was fine. The next day he arrived to sign the now-beautifully-bound document. Again he gave it a polite but cursory look and signed the original and copies with a specially presented pen. Following the ceremony, just as he was about to leave he said, 'Some day—but not now—tell me what you have changed.'

This incident was written into Shell folklore and bestowed on KY a reputation for genius which he did little to dispel. Many years later I recalled the incident and asked how he knew that there had been a change. 'Bob, it was simple,' he said. 'On Thursday there were 234 pages; on Friday, 236. Anyway I didn't need their agreement, I had their word.'

Success in joint ventures is all about personal chemistry and an ability to listen and understand each other's point of view before agreeing the way forward. Early on, KY had accepted that Ike Nadler, an old Shell man, should join his team. Ike was another extraordinary person. Having escaped from Romania and the invading Germans in 1939, he made his way to the Black Sea and hijacked a fishing boat. He dodged around enemy ships and was eventually picked up, out of drinking water and close to death, near Cyprus. After he had recovered he joined the British Army in the Middle East, from where he returned to Shell. I had been to Haifa with Ike a few years previously to buy the Sierra Leone refinery, but that mission was aborted when a military coup changed the government there and politics went into meltdown. We had enjoyed each other's company, so renewing the relationship in Bangkok was a real plus.

In the meantime the marketing and distribution business was going well and the refurbishment of assets was under way. Shell's acquisition of Billiton had brought us a tin-mining business and a tin smelter in a joint venture with an American partner, Union Carbide. The concession had been given under the aegis of the exiled field marshal, so it too got caught up with student protests about the Vietnam conflict. It was very messy, particularly since Union Carbide was still reeling from the toxic-gas-leak disaster at Bhopal.

Praphot, our deputy chief executive, was an experienced man who could articulate the problems but, challenged by my finance man John van der Linden, could not always provide the solution. John had spent his career in the East. He was well liked though sometimes brutally candid. Much later he met George Soros, the American investor and, having beaten him at chess, was invited to be his Thai investment adviser. John accepted on condition that the moment Soros stopped taking his advice, the arrangement was over. When Soros's investments prospered and he wanted to cash in, John told him to wait another two years. But, attracted by better opportunities elsewhere, Soros took the money, so the deal was amicably over.

Both John and Praphot were solid supports in an increasingly tense environment. Student power was on the rise, and our joint-venture operation was near the top of the hate list. Union Carbide saw no danger and dismissed our assessments: this was a signed legal deal and governments do not renege on those. To their legal representative, Thailand was just another American state; he had no comprehension of the country's centuries of independence!

My PR manager and mentor Somjai, a larger-than-life character, arranged a dinner with a retired general who was close to the royals. Somjai disappeared early and I was left alone with our guest. He was articulate and had many questions about Africa, the way things were done there, and how I approached things. At first the conversation had a counselling feel to it, but then he changed tack. He began to talk about the philosophy of the East and *I Ching, The Book of Change*, a product of China's classical age and considered to be the world's oldest divination tool. The book uses 64 hexagrams to impart ancient wisdom in the modern world. It was a set of precepts lauded by Confucius and, as I subsequently saw, contained so much common sense and insight that I adopted it as a personal guide. The book is still here in my study.

We moved on from *I Ching* to the *Tao Te Ching* which, he said, described Thailand. The author's philosophy was a recognition of forces, likening human affairs to streams of water that cannot be dammed because their power is irresistible, but can be guided because their destination is not.

We talked for a long time and as he left he put his hand on my arm and said, 'Thailand will change you—let it.' With that, he slipped away into the bright lights and hubbub of the Bangkok night. I never saw him again but he had made his mark. As I sat with my 64 hexagrams and my *Tao*, trying to restructure my approach to the complexities of life and business in Thailand, I felt it had been a watershed experience.

*   *   *   *   *

The increase in crude-oil prices had created mayhem in retail markets. For Thailand especially this was a bitter blow. Unlike many other things in that country, the starkness of the price rise could not be massaged into acceptability, and as the primary retailer we were the purveyor of bad news. Political help was not to hand. I told my colleagues that this was just the beginning; there would be further price adjustments and the upward trajectory was inevitable.

Statements of government revenue and expenditure in the oil-producing states are some of the best price indicators. If you analyse them, you can see which countries are in negative territory, and which are in positive. When expenditure became negative, OPEC tilted towards price action.

My father, third from right. The first time he ran away to join the army he was brought home: at 15 he was under age. Three months later he left again and 'kissed the Duchess's shilling' in Aberdeen, so enlisting in the Gordon Highlanders. By the end of the First World War he had served in France (including at the Somme), Turkey, and Italy. Still, he said, by far the worst moment of his life came when he had to tell my mother what had happened to me in the shop.

Back row, centre, in my final year at Castle Hill Primary School. My fellow left-hander, Sandy Merilees, is last-but-one in the same row.

My mother was keen to have a formal record of my golfing silverware. I agreed to pose for the shot aged 16—in part as consolation for the anxiety and distress that my escapades had already cost her.

Front row, fifth from left: head boy at Bell Baxter School.

Above left: Typically fine weather didn't deter the Scottish Boys' Championship crowd at North Berwick Golf Club in 1950.

Above right: *Golf Illustrated* used to invite readers to submit photographs of themselves for a professional critique of their technique. Two chums sent in this one of me, taken at St Andrews. It was returned with the note: 'Player should hold more tightly with his right hand.'

Crowds still throng St Andrews every year to watch 'Kate' and her uncle, college founder Bishop Kennedy, ride through the streets. The identity of the student who is to play Kate remains secret until the procession starts. To my astonishment, I was elected to the role in my first year. In my fourth I sat in as the bishop.

My mother, Elizabeth Margaret, lost her father in the 1917 flu pandemic. She worked as a clerk before marrying in 1932, then ran the business accounts through the difficult war and post-war years. My brother and I then both moved away: Jim west to Canada, and me east to Brunei and beyond.

Because we were to marry in Singapore, far from family and old friends, we advertised locally to find wedding guests who had links with home. Katherine Tracy (far left), whose father was Cupar's dentist, was one of many respondents who joined the celebrations.

Concerned about the health of the many unmarried mothers and their children in Port Harcourt, Jo (front row, third from left), helped to set up a babies' home. Run and managed by a joint Nigerian and expatriate team, it offered a supportive start to youngsters from the slums.

Above: Another night,
another party ... Jo and
friends gather to discuss
matters of moment in
Port Harcourt.

Left: Dancing with a
partner was also dancing on
your own: the rhythmic high
life at 9 Waring Road, our
Lagos home.

Siblings joined in at the Port Harcourt Primary end-of term party. Douglas (centre right) seems to have misread the dress code.

Jo and friends exchange ideas on local welfare initiatives.

Above: Lunch with (right) installation manager Ade Onafowokan, and (left) personnel director Mobolaji Ajose-Adeogun. A true friend in distressing times, Ajose-Adeogun subsequently became the first minister of the Federal Capital Territory of Nigeria, home of the country's new capital city Ajuba.

Left: Having befriended members of the Thai royal family at school in Cheltenham, Jo renewed the acquaintance in Bangkok. When the two younger princesses spent an evening at our home, Princess Sirindhorn treated us to a piano recital. Princess Chulabhorn (right) is King Bhumibol Adulyadej and Queen Sirikit's youngest daughter.

Mr Chow Chowkwanyun, KY to his many friends, was head of the Thai oil refinery company in partnership with Shell. A man of energy and bold ideas, he was great fun to work with. No KY enterprise was ever dull. Our friendship with him and his wife Sarunya (Lavinia) endured for decades.

Jo accepts an award from Thailand's Queen Sirikit. Her participation in the ceremony showed the extent to which she, a foreigner, was trusted and admired within the inner circles of the conservative elite.

Above left: The Reids enjoy a rare all-together moment on holiday in Bangkok.

Above right: Sir Garry Sobers lived for a time in Melbourne, his wife's home city. Cricketing kept him busy at weekends, but during the week he and I played golf. Ron Douglas, a former wartime pilot who now ran West Australia for Shell, would often join us for a relaxed threesome.

Nigel Mansell celebrates his first Formula One victory at the 1985 Shell Oils Grand Prix of Europe. I had been carefully rehearsed in presenting the heavy trophy one-handed, and was mightily relieved once it had been delivered intact.

Living in Australia, Land of Sport, I could hardly pass up the chance to compete at Royal Melbourne, one of the world's greatest golf courses. This image, taken at the 1981 Pro-Am Open, found its way into the 'Hall of Fame' at St Andrews University, where some decades earlier I had been pleased to achieve a golfing blue.

Opening the Tern and Eider North Sea oil platforms, The Queen Mother said, 'With my own home at the Castle of Mey on the northern coast of Caithness, I am probably your nearest neighbour. I send my greetings and good wishes to you all, not least to your family at home, whose courage and forbearance is such a splendid support to those working far out in the North Sea.'

Above: I have been fortunate to meet the Queen a number of times. Jo and I were invited to stay with her at Windsor Castle on one occasion. This was at the Albert Hall on the evening of a Royal Concert, which Shell would have been helping to support.

Below left: The idea behind a highly productive 'exchange programme' between oil workers and coal miners was to look at each other's operations, understand their respective challenges and perhaps share solutions. On the return visit, men from this mine in Merthyr, South Wales visited our offshore platforms in the North Sea.

Below right: Enjoying a joke with Ugandan-born lawyer Dr John Sentamu, who came to England after a difference of opinion with then-President Idi Amin. When John was Bishop of Stepney, Jo and he worked together on the Urban Learning Foundation to improve local facilities for trainee teachers. In 2005 John was appointed Archbishop of York.

Above left: At the beginning of my career with Shell, spin and sound-bite were alien concepts. Less so by this stage, when cameras and microphones seemed to be standing by to record every unguarded expression or remark. I learnt to be ready for them.

Above right: At the Channel Tunnel connection site a young supporter joins me for an overview of proceedings.

Below: Standing by at Waterloo for the train that would take Queen Elizabeth II and Prince Philip under the Channel to France. Her Majesty and President Mitterrand formally opened the Tunnel in May 1994.

Above: A grand day out: Bill McAlpine, grandson of 'Concrete Bob' Sir Robert, took Jo and me for a train ride in his garden. As an apprentice engineer in Hayes, Middlesex, where the company's wagon fleet was kept, Bill learned that its last steam locomotive, Hudswell Clarke 0-6-0ST No 31, was going to be scrapped. He had it transferred to his home at Fawley Hill, bought a section of track, and constructed a private railway almost a mile long.

Left: Sir Malcolm Rifkin (on Jo's left), then Secretary of State for Transport, seems focused elsewhere as I introduce The Queen to Chris Green, BR's director of London & South East. Chris was later appointed managing director of InterCity.

Above: The Great Western Society invited Jo and me to take a trip on the venerable locomotive *Nunney Castle* which, after 12 years in a Welsh scrapyard, had been restored and returned to steam in 1990.

Below left: John Cameron, chairman of BR's Scottish board, surprised us with a rail-retirement party at his home in Fife. Guests included old school friends who represented the local farming community and Scottish Railways.

Below right: The Princess Royal became BAFTA's president in 1973 and keenly supported Shell's efforts keep the organisation's spending under control. By the time of this awards ceremony 20 years later, its finances were keeping pace with its legendary sense of style.

Sir Bob (right) and his wife Joan and Mr Cameron with the loco-style nameplate.

arty for Sir Bob's retiral

Despite the hectic life we led we always made time for each other. We also followed my mother's rule never to go to sleep on a quarrel. It worked. I think this was on the last hole at Gleneagles. It had been a lovely careless afternoon and the weekend ahead looked good—but who knows in the oil business? We just felt blessed to be alone and together.

Douglas said 'Everyone held in their minds an image of Jo in full voice, with a joke, or a fiercely held opinion, or an insight, or a point of view—quite often a contrary point of view, but always a memorable and colourful one. We all had our pieces of that jigsaw which made up the larger-than-life picture that was Jo Reid. Just like those madcap jigsaws she insisted we did every Christmas.'

This is because in many OPEC countries, in order to maintain control of unsettled political situations, governments had to show their muscle. Nigeria, for example, which I knew well, was only one of the price hawks, but it had a burgeoning population and a programme of reconstruction and development that swallowed money. We needed to prepare all levels of government for more rises, and work towards a flexible price structure that would allow crude-oil price escalation to flow through into the retail and wholesale markets. Our refineries had to have the crude if the market was to be supplied. They operated on a processing-fee basis, so we had to purchase and process. In simple terms, the buck stopped with us.

Civil servants in Thailand were well travelled, well-educated and—according to my Dutch colleague John van der Linden—hopeless. John was at times less than subtle. After he retired he became editor (of sorts) of the *Bangkok Post*, and his wise leaders made compulsive reading, but his heavily biased analysis was at times flawed. Wimol, the key civil servant who had delivered the tin agreement for us after much hard work, had been a senior member of the Free Thai Movement during the war. He had parachuted into Thailand just as the Japanese retreated. John told the story that he had brought gold with him and, when he landed, chose a spot to bury it. He had then got lost in the jungle and never been able to find the gold again. John claimed Wimol wanted to return to the sky and jump again to try to find the place! Someday someone will be lucky.

Wimol's first lieutenant was Sivavong, a highly educated, well-travelled man who understood political reality. He was sufficiently outspoken to put the message across, and we kept him briefed on OPEC developments. The pair arranged an interview with Deputy Prime Minister Major General Pramarn Adireksarn, to talk about the oil situation. Praphot was my minder. He observed that Pramarn was the most serious and senior member of the political hierarchy and the military. When making a point to such a person in Thailand, he cautioned, you must act like a butterfly: alight gently, do not dwell and do not return to a subject. Move on instead to a quieter, more beautiful flower … such as golf!

We were received with great courtesy by the major general and after a few exchanges, mostly in Thai, I explained the problem in non-adversarial terms. I used Nigeria as an example, and he was surprised to hear that I had spent 13 years there. Having delivered our message and added a brief note about a proposed refinery expansion, we touched on golf, finished our tea, and left. Back at the office I asked Praphot, 'Was that like a butterfly?' He smiled enigmatically.

'Almost,' he said. Comparing that conversation with my unconstrained exchanges with the Nigerian ministry of mines and power reminded me of the stark difference between the two cultures.

Somjai, as always on the alert, asked how it went and we told him. 'I shall now brief the ministry,' he said. Praphot looked apprehensive and slipped into Thai. I changed the subject.

'He is keen on his golf,' I said. 'Why don't we give him one of those new centre-shaft putters?'

'Never!' Somjai exclaimed. 'How many putts do you think he misses? Do you want him to think of you every time a shot goes astray?' This was sound advice. Putter or not, though, and somewhat to my surprise, we managed to convey the inevitability of price escalation and secure a more flexible approach to accommodating it.

A change at the top of the government saw Admiral Sanghad take over. Sanghad was as unlike the quiet, courteous Pramarn as John Major was unlike Margaret Thatcher. KY Chow welcomed the change. 'He likes boats,' he said, 'and he likes sailing.' The ever-present Ike rolled his eyes and muttered, 'Chow, he is an *admiral*; please stop addressing him as *Animal* Sanghad!' For KY, with his heavy accent and ebullient, effervescent way of speaking, this was impossible.

Democracy had been given a short break to refresh itself (actually not a bad idea for other situations I could think of), and we went to see the bluff admiral. He was like the Nigerian Admiral Wey in many ways, except the gin was not as heavy. He told us that he wanted Thai children to participate in the world sailing competition in Turkey. Later we helped put together a small fleet of tiny single-sail Optimist dinghies, and after an intensive training programme the team set off. One little boy from the seaside who had never been to Bangkok travelled on a Jumbo Jet to Ankara. There was pay-per-view television available on the plane, but by ingeniously deploying a selection of drinking straws, he secured the sound and then the picture without charge! He enjoyed his journey, much amused by the cartoons. When they arrived he got the hang of the course, accepted the cold water and the rough winds and won a bronze medal. We were all delighted and proud of his achievement. The following year was a triumphant success for sailing and, more importantly, for tourism.

*   *   *   *   *

While things move slowly—and often sideways—in Thailand, this was not KY's MO. He saw the congested shipping and super-tanker movements around Singapore as an accident waiting to happen. There were plenty of precedents, the worst being the explosion of SS *Mont Blanc*, a ship carrying high explosives in the Halifax Harbour in Nova Scotia, Canada in 1917. It was involved in a collision at low speed and the resulting explosion obliterated every structure within a half-mile radius, killing some 2,000 people. The surviving residents of Halifax carried the scar of that disaster for a lifetime. So KY revived the idea of a canal across the Malaysian peninsula—the Kra canal. It was an interesting concept not just because the canal would be less congested and therefore safer, but also because the resulting shorter journey would be more economical. As the proposal began to gain traction, KY put forward the idea that it would be quicker and cheaper to put in place if a nuclear trenching instrument were developed to build it. A series of controlled explosions could quickly expose the route to the depth required.

KY was not a popular figure with the students, despite his interest in their education and support of their institutions, and this daring idea did nothing to increase his popularity. Indeed, quite the reverse: it sent them into orbit. KY's touch could be inspirational, but it made no concession to the audience. I do not think he ever understood the outcry or even registered it. It took all Ike's magic to close that chapter and move on.

The refinery prospered and we talked to ministers about our proposal for its expansion, the initiative that had defeated my tolerant and talented predecessor, Henni de Ruiter. He had got it in to the investment board, which was an achievement in itself, but not out. The proposal linked expansion to an extension of the refinery lease. But the public and press saw this as endorsing a decision made by the earlier military government and the ousted Field Marshal Thanom Kittikachorn, so it fell into the 'too difficult' box and the decision was deferred.

Kukrit Pramoj was prime minister. An Oxford graduate like his brother Seni, his wit and sense of humour were legendary. We attended a farewell dinner at his house in honour of the British ambassador, Sir David Cole. Kukrit gave a fine speech praising Sir David for his contribution during what had become a very difficult time in the region. Sir David made a typically modest reply, at which point Kukrit signalled for the British national anthem to be played. After the presentation, Kukrit turned to the French ambassador and thanked him also for his attendance, at which point to the great amusement of everyone but the honoured guest, the small orchestra burst into a lively rendition of Offenbach's *Galop Infernal*—the cancan!

During the dinner I mentioned that Frank—now Lord—McFadzean was due to visit, and that we would welcome an opportunity to pay the prime minister a courtesy call, so a few weeks later we met again. Lord M was always at home in Thailand, and at the outset he asked Kukrit why the Thai economy was so vigorous. 'It's simple,' Kukrit replied. 'We politicians interfere for only a few hours a day. It's in the rest of the day that progress is made.' We raised the issue of the refinery expansion and he asked me where we had got to.

'It's with the Board of Investment, Prime Minister,' I replied. He gave a great sigh and said,

'Within that confused bureaucracy it will sink without a trace.' I took the opportunity then to say I would write to him.

Somjai, keen to look after our visitor during the rest of his stay, made it his business to keep an eye on Lord M. Being a true Far Easterner, he loved a late-evening ramble through the shops and restaurants around his hotel. After the first evening, Somjai came to see me in worried mode: Frank had taken to wandering around with his wallet sticking out of his back pocket. Would I speak to him? We had no desire to tell Frank that he was being minded, so instead we sketched out to him the kind of crimes that were not unusual around the hotel, taking care not to malign what was mostly a quiet area. This did the trick and on the next walkabout at least, the wallet was safely tucked out of sight.

Thailand's unique reputation for indulging its visitors has helped build its huge tourism business. Like all cities, Bangkok had its sleazy side; this developed during the Vietnam conflict when it became a focus for troops seeking rest and recreation on leave. One of our visitors wished to experience the recreation part, so a member of my staff offered to take him out for a meal followed by a postprandial massage. He was shy when the moment came but he had his soapy shower, lay on a luxurious towel and supple fingers relieved his taut muscles. Beginning to relax he stretched out his arm but, as his fingers brushed the curtain, to his astonishment someone caught his hand and put a whisky soda in it. The idea that there was an audience in the room terrified him so much that he leapt from the couch and rushed for his clothes. My man had to reassure the masseuse that she had done nothing wrong, and silver changed hands. This buffoon, normally a serious man, left his watch behind in his haste and by breakfast with us next morning had given it up. This we put right—it was waiting to be collected from his bewildered masseuse!

Shortly after our time with Kukrit he lost power and the mantle passed to his brother Seni. He was our lawyer: a gentle, devout man ill-suited to the machinations and corruption of politics. Not long after his accession a territorial issue returned with a vengeance. The size of the acreage we had acquired in good faith for our offshore mine had been miscalculated—it exceeded by a minute proportion the maximum territorial lease allowable. The deal was declared illegal, so we would be required to surrender all the land. I heard this at midnight, and by seven o'clock the next morning Somjai had Praphot and me in a meeting with Seni.

First we asked him to confirm that the Cabinet decision was irrevocable. It was. 'Why don't we enter a joint venture?' he asked. I explained that this required shareholder approval, and even if we were quick we were not the only shareholder in the mining venture. But more fundamentally I explained there was the issue of investment in refinery expansion and possible exploration and production. Our intention in Shell Thailand was to give our shareholders confidence that the flow of investment would continue. An equity grab—because this was what was being proposed—would undermine our efforts. We explored other avenues that might allow the tin mining to continue while the territorial infringement was corrected, and there seemed to be some flexibility here.

Within a few weeks Union Carbide had sold its shares to us, giving us free rein. Billiton was flexible, and with the help of its chairman, Cor Herkstroter, we forged a deal in which we would be the operator, paid in tin, and tantalum (metal for hardening turbine blades) would be our bonus. This was a much-improved deal, and we now had full control of the management. So our portfolio was enhanced and our prosperity with it. It did nothing for our refinery aspirations, although maintaining the investment commitment did help speed the conclusion of the mining deal.

This commitment had become very important in what was seen to be a deteriorating political situation in the Far East. The Vietnam conflict was bleeding to its conclusion, communism was hardening in Laos, and Cambodia was in the final stages of collapse with Pnom Penh surrounded and under constant barrage. Radical developments were imminent and they would affect the uncertain political situation in Thailand.

*   *   *   *   *

Northern Thailand is in the middle of the 'Golden Triangle'. Some believe that the remains of the Chinese Imperial Army, which lost the war to

Mao Tse Tung's communist army, still exist inside this remote wild area where the heroin poppy is grown. Attempts to substitute the crop failed for a variety of reasons, among them the drug lords' coercive power, the attraction of ready cash and the well-known unreliability of government promises. And so the white stuff is there, its routes to market protected and any obstacles removed; in this world of shadows, nothing is clear but the impossibility of interrupting the vicious trade.

During one of our expeditions to explore the country, Jo and I went to the Cameron Highlands on our way to Penang. Some time earlier, with one of my predecessors at Shell, I had met Connie Mangskau here. An attractive, vivacious woman, she had accompanied an American ex-military man, Jim Thompson, on holiday to the Highlands in March 1967. Jim had been trained by the United States Office of Strategic Services (OSS) and was assigned in 1945 to liberate Thailand from the Japanese just as Japan surrendered.

He settled in Bangkok after the war, integrating easily into Bangkok and Thai society, and made some commercial investments. The rumour that he continued as an OSS intelligence chief is lent credence by his circle of Laotian and Cambodian friends, all of whom were reformists by nature and zeal. Such an infiltration to forestall the communist advance would appear logical, considering the developments at the time.

The Thai government's inclination for neutrality and its aversion to any political involvement that might damage its sovereignty made Thompson unpopular with the establishment, (which also jealously coveted his increasingly valuable collection of Thai artefacts). But then, rather like the Thai national anthem which never seems to end, there is always something else. The Americans needed a staging post for their flights over Vietnam, and Korat in northern Thailand was an ideal choice. Just as this location was being prepared, Thompson, a cultured individual with a highly developed sense of taste and fashion, had set to work developing the Thai silk industry which had hitherto been based in the rural north. It was a grass-roots operation so Jim was popular with Thai farmers and fêted whenever he visited.

Against this colourful background, Thompson took a holiday trip with Connie to the Highlands. One Sunday morning he set out on a walk, never to return. His disappearance generated one of the largest land searches in Southeast Asian history. The Dyak trackers, my friends from the jungles of Sarawak and the ultimate professionals, were called in, but all they could find was that the trail went cold in the jungle not far from where he had set out. The mystery remained, only compounded by the murder of Jim's sister in New York some months later. Neither the disappearance nor the murder are closed cases.

In Africa the dangerous war had been a civil one conducted within the state. The opponents were easily recognized by tribe, as were their motivators. In the Thai situation the dangers were external. The domino theory was all about rearrangement of the power blocs, but its weakness was that it ignored two unknowns. First, were the Vietcong going to act against their neighbour? After a long, draining war did they have the stomach or desire to go on and take someone else's troubled country? Second, would the Chinese sit by and see the balance of power change and a stronger and potentially adversarial bloc emerge to challenge their own position? Kukrit had strong views on this. He did not think China would let the process get out of hand and this, he told us, he had had confirmed from the very top. China after all, he said, was the suzerain.

*　　*　　*　　*　　*

Paul, our middle son, had been swimming near the rocks when he stood on a sea urchin. His howls were loud enough to waken the dead. Jimmy, one of my younger colleagues, advised us not to try to pull out the needles because they would break, leaving their poison alive in the foot. The remedy, he said, lay in neutralizing them and then battering them to death. Neutralization, however, called for a bottle of gin, which we didn't have to hand. Jimmy suggested we break into a colleague's nearby beach house to obtain the necessary, which he did, and emerged triumphant, having recompensed the outraged steward.

Paul lay there intrigued, quieter now that the pain barrier had been passed, and watched the needles, swimming in gin, slowly release their hold. Afterwards, while we finished the gin, we got talking about the absent donor.

Kevin, or 'Duck Feathers' as he was known, had been interested in Vietnam before he returned to Thailand. He was not a Jim Thompson, but he had made Thailand his home. He had built his business harvesting duck feathers from the butchers and farmers who prepared carcasses for market in Bangkok, Singapore and Hong Kong. He processed the feathers for pillows and duvets which he sold on the domestic and export markets. This was a good and simple story. If it, too, was a cover, it was a good one!

*   *   *   *   *

The news came almost out of the blue: the Americans were out of Vietnam. Their air bases were closed in Thailand. In a single week they were unceremoniously expatriated—and then, less than a fortnight later, a new alliance was struck. In a typical Thai move, the status quo was broken and restored. The predictors of doom and domino theorists had got it wrong, having underestimated the political expertise that had kept Thailand secure for centuries. So much newsprint and intellectual brain-power wasted on another redundant Western supposition!

Laos, however, was a different tale. It had been a sleepy kingdom holding a strategically-placed enclave. It was bordered by the Mekong which fed from Thailand and had to be crossed by ferry. Every day, oil and foodstuffs made their way across this river, which also watered the Mekong Delta paddy fields. The stream was of such economic and political importance that the King of Thailand himself was in charge of the water resources department.

Laos's traditional government collapsed under the pressure of a virile and at times virulent young Communist party, after which the government's bizarre doctrine of 'No alternative' tolerated us. So we stayed. The supply bills were paid and service fees were nodded through. It was not the economics that worried us; it was our staff's circumstances, the fearful atmosphere in which senior management lived. People were being uprooted from their homes for no apparent reason and taken to isolated camps on islets on the Mekong.

Our first objective was to remove our senior Laotian management to safety, but this was easier said than done. We organized a series of table-tennis matches between them and their neighbours in Bangkok and secured some exits that way: visitors made up the numbers when the teams returned. Somehow in this covert evacuation, a tiny orphan managed to get herself included. Years later the young woman legitimised her departure, but not before she had taken an honours degree in classics. She was our resident communist for a while in Bangkok, and we missed her when she went.

I was increasingly concerned that staff's health was suffering because of deficiencies in their diet. Since we had land aplenty around our depot, we began quietly cultivating crops, and then added a chicken farm. This engaged staff and spouses alike and our farming successes and failures were duly noted on the weekly reports. Morale was greatly helped by two young people, a serious manager, Tim Faithfull, and a colourful Scotsman, James Davidson. The stability of the one coupled with the effervescence of the other did much to keep the corporation in a healthy, positive state in a deteriorating environment.

In a way visiting Laos was like my old practice of visiting Uganda, except that in the former one was more likely to be incarcerated than killed. Against this gloomy background we decided to hold an AGM to confirm our capitalist status and create an upbeat diversion. We signed the accounts and then proceeded to the important business: the entertainment. Here again my experience of Sunday School parties came into play. After the food had been devoured with relish, the games started with pass the parcel. The music had been tricky to compile because Western pop music was out, so James Davidson had concocted a hash of national anthems with chunks of Berlioz and a few hymns thrown in. The last person to hold the parcel in a dramatic play-off won the chicken. The second game, musical chairs, was another winner; as the chairs were removed, people ended up on one another's knees and laughter could be heard streets away. Excitement had reached fever pitch by the last game, the statue waltz. In this one you must freeze and stay absolutely still when the music stops; if you wobble, you're out.

By this time the soldiers patrolling the premises had graduated from being alarmed to being intrigued, particularly when they saw people standing motionless in silence. Putting their rifles to one side, they joined in. One of them won a capitalist prize of a box of eggs for an exceptional performance: hours of standing to attention had clearly prepared him well. Visiting Russian commissars, most of whom seemed to wander aimlessly through the streets making little attempt to engage with their diminutive communist brothers, looked on at this strange congregation, shaking their heads at the abandoned rifles and helmets and the erratic blaring of music.

The AGM I believe continues to this day, its origins probably lost in the mists of time. In the context of my early conversation with my Thai sage I looked at Laos and thought about that passage in the *Tao Te Ching*:

> Yield and overcome;
> Bend and be straight;
> Empty and be full;
> Wear out and be new;
> Have little and gain;
> Have much and be confused.

It reflected the struggle and achievement of our tiny family in Laos, and the distressed confusion which, from time to time, upset our wealthy neighbours.

*   *   *   *   *

The Dutch Ambassador's National Day party underlined my view that while things were immobilized in the West, the East was changing. Something seemed to have affected China's decision-making oligarchs—the prosperity of its neighbours, perhaps, or the market opportunities, or the change of leadership and reformers' determination. Whatever it was, it was significant that the Chinese ambassador attended the party not in a Mao suit but in a Western suit. What is more, he talked openly about his country, its development and future. The visiting director from The Hague was amazed by this demonstration of a new China and told his fellow directors. This was one of the first indications of a shift in direction, and it presaged exciting potential.

For Thailand it was business as usual. The tourist boom quickly filled the vacuum left by the departing troops. As part of the tourist push, the World Orchid Festival was organized, to be staged in the rose garden at the national cultural centre. Thai forests are filled with many exquisite orchid varieties, and even today plant-hunters dream of discovering new ones among them.

We had two visitors at the time who could not have been more different from each other. One was a distinguished Indian refinery executive who cultivated orchids commercially. We could imagine him, immaculately be-suited, inspecting his orchid groves from his air-conditioned Land Cruiser. Our other visitor was a professor who was determined to go up country to collect species in their natural habitat. He knew where he wanted to go and it was exactly where Somjai did not want him to go—into the middle of the Golden Triangle, where agents who control the processing and trafficking of opium lie.

But our professor was persistent. He had not come to the festival for lunches and dinners, he said; he wanted to be off foraging for new species—or something else. We obtained permits and hired a jeep, and he sped away from the disapproving Somjai, for whom each day he was gone was a day filled with dread. But after a fortnight our errant professor returned loaded with plants and stories of his adventures. Somjai believed the plants, but not the stories. In truth I was never sure.

In a couple of days the professor had gone, and his plants followed. Just a few weeks later a barge sank in the Chao Bay, only to be quickly surrounded by police boats—and then silence. No more news of barge or cargo. The white stuff is like water: its flow is unstoppable until man no longer wants to drink.

The Vietnamese conflict was over and the elderly Chinese ambassador in Laos asked me to visit him. His low number on the Chinese membership role confirmed his standing and access to power. After a few pleasantries, he said, 'You have concessions off Vietnam shores and we wonder when you plan to develop them.' This was a volte-face almost as complete as Thailand's expulsion and welcome back of the Americans. I advised him that as these concessions were owned by Shell Oil, a public company in the United States, I would route his enquiry immediately to its headquarters in Houston, Texas. I also warned him that it might take some time to elicit a positive decision, since wartime regulations regarding trade would not have been revised and emotions understandably ran high in America. As it turned out, I was not wrong!

*   *   *   *   *

When the Easter holidays arrived, we set off with our Thai friends for a celebratory jaunt. Our party ran the supply chain: duck soup, noodles with shredded chicken, peppered cashews, hard-boiled eggs and fruits— rambutans, mangosteens, lychees and mangoes. The mangoes ranged from delicious voluptuous fruits to bitter green varieties which, with a glass of sweet lemon water, refreshed you and relieved the heat. And the sit-down meal was yet to come!

Before we set off up Khao Yai in the western part of the Sankamphaeng mountain range, we stopped at the service station. Our dealer there had made a tremendous business out of selling small bottles of upper-cylinder lubricant to those facing the climb. It looked to me like a straightforward hill, but the sales pitch went as follows: 'While you prepare yourself for the climb ahead, prepare your engine. While you spice your food, give your engine some *ajinomoto* with its petrol!' Sales of the little bottles grew and we all got up the hill a little wealthier.

Tigers (in groups they're known as 'streaks') lived in Khao Yai national park. We left the car at a rest house and, as dusk fell, set off to look for them. Douglas, then 17 and selected by Somjai for his air of responsibility, was handed the searchlight and, to his great joy, a loaded Smith & Wesson revolver—to be used only in the event of attack. This added to the drama, particularly for the Thais who saw in Douglas a natural tiger-hunter, not a security man. But no heroics were called for. No tigers to shoot.

Gin rummy followed supper as the temperature fell. According to the thermometer hung outside, it fell rapidly—possibly because Paul had slipped out to wrap ice around it. Somjai checked it and came in to declare unheard-of lows, while Paul surreptitiously reapplied the ice bath. So it went on until all our Thai ladies were smothered in blankets, and all the men were huddled in jerseys and golf jackets. The episode endeared Paul to Somjai for life.

Golf the next day was on the hills around Khao Yai. There were guards dotted about us as we played a fine course with spectacular views. Close to the 12th hole there was a blood-curdling cry as Somjai and his partners fled towards the clubhouse. We played on quietly—until we too reached the tiger footprints. Was that Paul Reid again, playing tiger? Those who did not want to stay and find out hot-footed it to safety; the rest of us completed a delightfully tranquil round.

The Thais' ability to enjoy themselves is well known. Despite all their problems, they made the most of their situation and brought out the best in their friends. This was a happy period of our life and any reunion brings it all back.

*   *   *   *   *

Shell had been pleased to sell KY a couple of tankers to underpin his supply position. He named one of them *Siam* after his daughter, and the other *Mona* after his granddaughter. The blessing ceremony was Buddhist, and the captain had a near-heart attack when he saw the priest lighting joss sticks on the deck above the oil tanks.

Any function KY was involved in was filled with the unexpected. We travelled together to London to check arrangements for the annual cocktail party. On our first evening we planned a quiet dinner with KY's daughter, who had just achieved first-class honours in physiology. In one of his announcement moments, KY said he had arranged medical studies for her at a top London hospital. This was like the Kra canal, only worse—she had already arranged to pursue a career in acupuncture! Her studies had prepared her well for it, and in a way it was a return to her Chinese roots. KY exploded: she was pursuing black magic! This was witchcraft! And so it went on. The food came and went, untouched.

The row gradually subsided and father and daughter relaxed to fight another day—but not the next day. That was company business. 'This place is too hot,' said KY as soon as we arrived to check the party venue. 'I shall take action!' And so he did. He hired what can only be described as whirlwind fans, and when the guests arrived the following evening he gave the signal to switch on. As the fans accelerated through the gears to top speed, dresses billowed above waistlines and hair, skirts and scarves all streamed northward. KY was somewhat short-sighted so was slow to view either the predicament or the lingerie. This was what operators are for and at the throw of a switch, order and decorum were restored.

*   *   *   *   *

Dredging off Phuket was not an easy business. Unlicensed miners picked up tin-bearing cassiterite from the seabed using garage hoses as pumps. Having been brought to the surface, it was illegally shipped by fast boat to Penang in Malaysia, where it was smelted. One day a tropical storm of typhoon proportions wiped out dozens of illegal miners at a stroke; fishermen were also caught up in the disaster. In an industry involved in prising resources from the earth, we are all in the same family. We all face the same catastrophic risk and I sorrowed for the men as I did for Mansor bin Kipli, lost off that capsized platform in the China Sea.

Thai funeral ceremonies have a dignity, a gravitas about them. Much later a great golfing friend of mine was killed in a crash flying into the Phuket area in a lightning storm. He was married to a Thai woman and had spent most of his adult life in Thailand. I was among Thai and *farang* (expatriate) friends at that funeral, and as we filed past the pyre, the Thai Airways staff in full uniform stood to attention. The courage and solemnity with which they faced the family and the mourners were very moving. As I paid my respects, I found life hard to understand. Somehow the frigid cold of an Aberdeen church seemed more appropriate to death than a warm garden filled with hibiscus and frangipani.

Later that month KY was in festive mood, telling our boys that this was winter—even although the temperature was in the 80s. We had to eat dumplings, northern Chinese food, preceded by hot-and-sour soup. With a term of school food behind them, they swallowed the lot. And then KY's stories started. First, one about poverty—a bowl of rice and a picture of a fish on the wall. Then his first insurance venture, insuring his school class against failure. Paul, to KY's delight, described this as a 'no-brainer'. And so the evening went on. Thailand had a way of forgetting its problems, and the food and the alcohol helped.

In the Songkran or New Year Water Festival, everybody ends up soaked, splashed by respectable—or those you thought were respectable—citizens wielding whatever water-carrier or -shooter comes to hand. But this celebration was over when suddenly our office building started to shake. Some thought they were having a heart attack, others braced themselves for another shock. Floor marshals led people down the stairs. I set about checking everyone had left—but Van der Linden was still at his desk, saying, 'Why do you think we leased this building? It can withstand anything on the Richter scale ....' That didn't convince the staff, so the next day the monks blessed the building yet again, and we all trooped back in, reassured. By this time I was beginning to believe!

*   *   *   *   *

Our neighbouring countries had calmed down a little by the time a report came through that the Baader-Meinhof Gang had split and some of its members were offshore going east, possibly to Bangkok. Being internationals we were considered potential targets, so an official arrived to make a risk assessment. He tracked me for a week, after which he concluded that my movements were too erratic for me to be singled out, and half the time I was not in a car (a favourite target) anyway. The refinery for some reason he found too open, too beautifully situated and too Thai to be vulnerable. As far as we know the gang or its members never came, their violent story continuing elsewhere.

The pirates of southern Thailand were more of a problem. They were hijacking our coastal shipping and looking for ransom. This was not a straightforward story and became even less so when someone proposed we should guarantee ransom. Van der Linden was apoplectic at the prospect. Praphot gave an enigmatic smile and shook his head. Sarisdiguna, a brilliant executive who became chairman of Thai operations, went to the navy ... and the problem duly vanished.

The marketing and distribution business was going well. The refinery was a financial success but the idea of its expansion had not progressed. The tin business was now a bonanza and the smelter was turning in good profits. At last we were poised to explore—but that adventure was to be for my successor; our time had come to an end. As we had spent four years in Thailand and in the most difficult times, Somjai and Sarisdiguna arranged for me to say farewell to the king at the summer palace in Hua Hin—a beautiful seaside town with a superb golf course designed by a Scottish railway engineer.

I presented myself at the palace, having been warned that my audience would be brief because the king was going out. I entered and bowed, but this slim, distinguished, keen-faced man waved deference away. He sat me down and asked about my stay. I said that with the troubles with our neighbours in Laos there had been difficult times for all of us, but we had come through it well. The business was sound and the future looked good. Our family had enjoyed Thailand, learned much and were grateful, and I thanked him.

He then asked me if I believed in planning. I replied that planning had to be based on reality, not on dreams. If it was based on reality, then you could invest with confidence. He said one of the problems in Thailand was getting people to look beyond Bangkok. It was in the rest of Thailand—in the villages and towns and the farms—that the future lay. Talking about communism, he said he could stay only if they, the people, wished him to stay. He then returned to his question of planning.

'I am responsible for water,' he said, which I knew. 'The officials came to me with a plan for a large new dam. At the conference table they laid out the maps and their plans. I knew the area well because in my youth I had covered the ground on foot. I said to them "This is a superb design and a magnificent dam. However, it is also unique."' He went on to explain to the nonplussed consultants and bureaucrats, 'It is unique because the water will flow up and not down into the dam.' He had called for helicopters and flown them to the area, where they duly tramped *up*hill to the proposed dam site. 'So you can understand why I am sceptical about planning,' he said. I could only agree.

The audience lasted over an hour. For me it was a rare experience to listen to one of the personalities of the twentieth century: a god-king who talked both humbly and knowledgeably about his kingdom and his people, and yet was open and realistic about the Realpolitik. My Thai assignment had been worth it just for this meeting.

At Jo's suggestion we stopped off in India on our way home and made our way to Gulmarg in the Himalayas. We stayed in Srinagar on a houseboat on the lake. It was spectacular as well as romantic, the gleaming mountains magically reflected in the water. Thailand had been nonstop for both of us and these few days of peace were restorative as well as offering a tranquil opportunity to review our Thai adventure.

*   *   *   *   *

Jo's interest in history had quickly led to a fascination with the evolution of Thai society. As a girl at school in Cheltenham, she had made friends with a member of the royal family, and this had serendipitously opened doors into Thai high society. During our stay friends had arranged for Jo to meet the Queen, which had been especially fascinating because the court members had also been there. One of the young princesses, a delightful quiet person, easy to talk to, came to a piano recital and supper at our home. Her sister was more evident in the public arena and had a following pushing her towards the throne.

There was a thriving farang society in Bangkok made up of old Far-Eastern hands. They populated the race and golf courses of the Bangkok sports club. Their Thai was often fluent and their contacts spread across all levels of society. One of the seriously wealthy members approached me with a proposition to join him and take over his business. I explained that my loyalty to Shell came first, but thanked him for his offer. Jo was in full agreement despite her enjoyment of city and the country. Her interests were increasingly elsewhere, back in her own land.

The assertion made early on by my mysterious dinner guest, 'Thailand will change you—let it!' was prescient. My ability to accommodate alternatives without deserting my principles had been strengthened by the unusual situations I'd been confronted with. Thailand had imbued in me a new calmness.

6

The Second Oil Crisis: Time to Trade

There is no way to approach a maelstrom—it just happens. Events aggregate in rapid confusion and collision is inevitable. To distil patterns from it is impossible; only time and exhaustion bring a return to some form of stability. But turmoil can throw up tremendous commercial opportunities, and that's where the excitement lies.

When I next returned to London, it was to work in the supply business. My first assignment was with the Group's aviation arm, which contracted airline business, coordinated prices from different locations and then monitored the payables and customer performance. It was well organized, the section leaders knowing their regions and customers well. Excellent and apparently inexhaustible engineering manager Terry O'Keefe oversaw the operation's technical integrity, the quality of the fuel and the safety of depots and fuelling. Joe Ludeman, with an encyclopaedic knowledge of his market and a sure touch, was in charge of the commercial side. The question for me was where I could add value without disturbing this smooth-running machine. My role had to be in enhancing the proceeds in what was a highly competitive market.

After I told the chairman that I had an evening job and could do with a day job, products trading was added to my portfolio. This was a more exciting role with plenty of opportunity to do deals and improve the bottom line. The Russian fuel-oil contract that we signed in a haze of vodka and Champagne was a good example.

As vice-president of products, I was on the trading company's management committee. The boss was a long-time central offices man with a fine record in the markets, but he was not easy to engage. At my first meeting, I asked a question that had been playing on my mind: 'How are we placed when the Shah goes?' His answer was brief and dismissive. Within a month, the royal reign of Shah Mohammad Reza Pahlavi collapsed and Sheikh Khomeini took control as grand ayatollah of the Islamic Republic. It was soon clear that the Iranians would cut production and we would have to fight for our share, since all contracts would be shredded in the icy wind of the new regime. Silvan Robinson, an intellectual, shrewd and vastly experienced man who became president of the trading company, took the lead. We travelled to Iran to see what we could do. Our relationships in Iran had always been good, so access was no problem, but progress was.

As we ate supper in the safe house, we reviewed our position. First, there was a queue of buyers swamping the offices; secondly, there was a terrified, if not paralysed, civil service, each member bracing himself for investigation and fearing a root-and-branch clear-out—or worse. There was little point in trying to pressurize these individuals, although we would certainly fight to keep our place in the queue. Thirdly, our hotel bedrooms were probably safer than this 'safe' house situated between two villas, guarded by active machine-gun posts. Fourthly (and critically), the cook was almost out of whisky and he needed to be financed. The price of whisky had risen even higher and faster than the price of oil. So we would have our meetings tomorrow and then decide who, in what was obviously going to be a long, drawn-out process, would come back and when.

Silvan's worldwide supply chain was already in chaos in this frenzied market. It looked like I would be doing the return Iran trip. The next day I sat in a meeting with the officials and a mullah, who listened attentively, but offered nothing. The meeting confirmed that our supply would be cut but we should continue to load our cargoes, and I expressed our appreciation for this. I then made the plea for more oil, but it fell on deaf ears. Silvan did marginally better with the big guns, showing that old friendships counted for something.

In a rather rash move we bought caviar and manoeuvred it through customs. At home it was well received, unlike our stark report. We were definitely going to be short, and it would take all our energy and skill to meet the deficit.

I decided to try Abu Dhabi, and in the guise of a courtesy call I visited the ministry of petroleum. The civil servant involved in crude-oil sales was an Iraqi whose family had been killed by Saddam Hussein, and their lands confiscated. He had been at university in the US when this happened, and after graduation he had made his way to Abu Dhabi. The courtesy call quickly became a business call, and we agreed terms for a sizeable purchase of crude oil and a price. Within two days we had a call from an Abu Dhabi prince in saying he could find us oil on identical terms plus a finder's fee. I showed the civil servant the message. 'Ignore it,' he said, without a moment's hesitation. I explained that I had to circulate my deals not only to the board but also to a list of princes, some of whom try to do their own business on top of it. 'They try it,' he said, 'because sometimes it works.' I ignored the prince, but our face-to-face deal secured a flow of oil through the crisis and an unexpected friendship without a hint of corruption.

In the meantime my able product man had come across a possibility in Greece. John Latsis, trader and ship owner, had care of oil in Rotterdam tanks, but it could be released only to an established refiner. As we made our way to Athens I learned about Latsis and his entrepreneurial activities in Saudi Arabia. He had facilitated the construction boom by using a cruise ship to house his workers. The flexibility of life on a licensed cruise ship could appeal to immigrant workers, and what was more it was easy to withdraw a force swiftly when the job was done. Latsis had been selected as a guardian of stored oil precisely because of the respect he had shown for Saudi Arabia in his activities there.

The first discussion took place with Dr Gregoriades, a distinguished figure fluent in English who always came up with an unexpected angle. The arrangement by which he distributed his products (a so-called hospitality agreement) had unfortunately just been cancelled by our Greek colleagues, one of whom was in the meeting with me. He had not mentioned this, far less briefed me, which left me somewhat at a loss when the good doctor enquired about it. I promised to look into it and tried to get back on track. I did not want to get involved in a domestic dispute, but if I could solve it, it might be the key to our crude-oil deal. My colleague reluctantly agreed— and I had a flicker of suspicion he might not have known anything about the cancellation in the first place.

I put my cards on the table: we came looking for a deal and found we had just removed one; did this stand in our way? Gregoriades paused and eventually said, 'No. Let's eat.' Over lunch we disclosed the volume (15 million barrels), and agreed the market price. They could deliver at established tanker rates, we would nominate the shipping programme and destinations, and they the ships. Each aspect had to be haggled over, and in virtually every area they had a point. Their being prepared to accept losses surprised our tanker people, who were also amazed at how the Latsis operatives squeezed the tankers dry. This in itself had long-term value for us.

The injection of Latsis oil made good some of our deficit but we were still short, so I returned to Iran. The Khomeini regime was now in full swing. Straight off the aeroplane I was confronted by men in raincoats who wordlessly studied my passport before handing it back. I thought there would be more of the same inside the building, but the immigration men had disappeared, as had the customs people. I plunged into the heaving mass in the arrivals hall, cashed some dollars and bought an excellent cup of coffee. Nobody paid the slightest attention to me—and there was I expecting a wave of hostility against the foreigner. In a few minutes our man turned up, and we were on our way.

Tehran is a beautiful city set against a range of distant mountains. It has an imperial feel about it and I always enjoyed coming back. The ministry had given me a number of appointments and as I worked through them it gradually became obvious they had a problem. What it was and where it was, I could not immediately see. I fished around and found it at last: it was the fuel oil. Their refinery tanks were rapidly filling up to the point at which they would have to curtail refinery runs, which in turn would affect their gasoline and gas oil supplies. With no secondary capacity, they would not be able to strip any more product out of the fuel oil, the bottom of the barrel.

Once again the haggling began. We would lift the fuel oil, but we needed an incentive, such as more crude. There was mileage in this. We set a date for my refinery man to visit and agree specifications, shipping arrangements and a programme. There was debate about how much crude we could have, but eventually a fair proportion of what we had lost in the initial cut-back was restored.

Then the fun began. René Guillerme, a French refiner, had joined my team. He visited the refinery, established the richness of the fuel oil and we began to line up customers who had capacity to extract the gas oil. We soon had more customers than product within the Group. Our business offer was to deliver the fuel oil at a price, allow the customer to keep the gas oil, then take the fuel oil back and on to Cuba to satisfy a Russian obligation, receiving in return aviation kerosene in India, which delighted my aviation chums. This was a dream deal. We then went to Japan to see the companies who had had to accept fuel oil if they wanted crude oil.

They initially thought we were sellers, and we did not disabuse them of the idea as the prices got lower and lower. When we reversed our position to become buyers, they were willing sellers—though very *un*willing at the price they themselves had set! But sell they did, again giving us a very attractive margin.

By the time this was all negotiated it was nearly Christmas, and Jo and I needed to get away from it all, so we booked a holiday on Islay to spend time with the family there. It was peace on earth in a land of gales and storms … until the phone rang. The Iranians wanted to see me. My farewell from Jo was icy as I climbed into the small plane that would take me off the island and into Glasgow, then to London and Tehran.

The deal was on the table for more crude and more fuel oil. After some tedious negotiation we were there, the deal signed and almost sealed. I returned home on New Year's Eve to an empty house and waited for Jo and the boys, who were making their way south after a stormy crossing on the seagoing ferry. Just as they arrived the phone rang again—the Iranians wanted me back. This is when you know you have a solid marriage. Not more than a few words passed—and there was certainly no fond farewell!

I have never worked out whether the immediate recall was a deliberate strategy or whether the need to complete and the pressure from upstairs was so great that the Iranians had no choice. My American director said I was the 'point man'; from my knowledge of cowboy films, that's the one who gets shot at. In the end it did not take long, and when it was done, the oil flowed. Jo's forbearance in this imbroglio gave her a place in the Shell history book. When the chairman, who knew her well, heard of her beyond-caustic reaction to their apologies, he assured the personnel director she was fine. She knew the way things were.

In the course of all this the Iranians had taken American hostages, which had obviously raised the stakes. My visit to the British embassy in Tehran confirmed that. There I found Arthur Wyatt, an acquaintance from Nigeria, now acting UK ambassador, chipping balls around the palm trees. I hit a few as we talked. He was spending his nights in a safe house and during the day did not have much to do—which, as I recall, was the case in Lagos. He was well trained for it. But Arthur was shrewd. 'It's dangerous,' he said. 'They know who you are.' On that happy note I left him chipping away and wondered where the signal to leave would come from, but it did not. Buying a new plane ticket in this land of *personae-non-gratae* Yankees was, strangely, possible only with American Express. My stocks of caviar were growing but my domestic supply of Scotch was falling.

Slowly our Iranian business settled down to a normal pattern. The Iranians were helpful and good to do business with, but always demanding. Why not? After all, it was their resource. I returned to them many years later to find that a good precedent had been set, which pleased me greatly.

*   *   *   *   *

The fast-rising cost of oil was now causing a pricing problem in my aviation business. The team wanted to defer inevitable price rises for as long as possible to maintain good customer relationships. I made two points. First, we had a low-margin business as it stood without oil-cost hikes; secondly the availability of fuel was not assured in any case, and customers had to understand that. We set about preparing a price that would cover the cost increases and restore a reasonable margin in this uncertain marketplace.

Our first encounter was with El Al, which showed little appreciation of our continuing to supply when others had deserted them. In response to our presentation, the company simply asked for a reduction in present pricing levels. They had never heard 'No' before or seen a 'Take it or leave it' deal, so we left them to discuss it among themselves. They subsequently accepted, reluctantly but with no bad grace.

Roma and Alitalia were next. Here our man was Dottore Arnaldo de Bernardis, an Italian with an effusive manner and a range of influential contacts. I had called in advance to brief him on the seriousness of our position and our substantial discrepancy on price. When he met me at the airport, he said all had been arranged. 'I have booked a very special hotel for you; you will like it.' Arnaldo's taste was never in doubt when it came to the best bits of living. It was late, so he took me to the hotel and introduced me to the manager, who seemed pretty active at 11 o'clock at night. Next morning I rose early and took a walk before breakfast. Wandering through Rome's architectural treasures as the city wakes is a pleasure not to be missed.

The breakfast room was virtually full—indeed, I took the last table. Looking around I saw that everyone in the room save the waiting staff was a cardinal. All were deep in conversation, which allowed me to study them one by one. Europeans, Africans and South Americans were all mixed up together and far too involved to notice this rather lapsed Presbyterian in their midst. When Arnaldo asked about the hotel later, I said it was a pity that the Pope hadn't joined us. Without even a smile he explained that His Eminence always eats in his own residence.

Alitalia staff were in full attendance. My slides were good and had been well tested, my presentation enlivened by anecdotes from Tehran and Abu Dhabi. The Italians were well aware of the supply crisis and the price escalation; you could not help feeling they had heard the same story from other sources. Afterwards we moved on to lunch: a compulsory event orchestrated by Arnaldo.

I asked if he felt the slides had worked. He was fulsome in his praise. 'They made the story and the point, now we have a profitable business again.' But I was concerned that the senior man had been wearing dark glasses throughout, and asked if that was usual. 'Yes,' said Arnaldo. 'He's blind.'

'But my slides ...?'

'He got the message.' And so to the job that Arnaldo did best: ordering food and wine.

Roma and Alitalia signed, but we had difficulties with the oil distribution at Fiumicino Airport. There we were dependent on Agip, the fuel-oil retailer, which could not get the lorries alongside the aircraft in good time. This meant more food and more wine and a promise to visit Arnaldo's Tuscan vineyard in a month's time. Until then supplies continued and he could sleep at night.

During that chaotic four weeks Jo and I set off for Warsaw to attend celebrations for the Polish airline Lot's 50 years of flying. We had been its solo supplier for almost all that time. To mark the relationship, Jo had acquired a beautiful crystal decanter as an anniversary present. Rightly she forbade any engraving and we added a dozen bottles of claret to help keep it full. On the plane our seats were 'honour' seats, which came with complimentary bottles of malt whisky.

We paid a courtesy call to Bill Forster, our old Nigerian colleague and his wife, Judith, the re-patriator of dead cooks. The welcome was as warm as ever on a very cold day. The official dinner was set for the evening. On arrival, Jo spotted the decanter on the gift table alongside some plastic helmets and a security jacket. Without a moment's hesitation she scooped it up and plonked it where it deserved to be: at the head of the table.

The speeches were laudatory and the Russians, easily recognizable by their uniforms, size and style, made a shot at congratulations. This was before the wall came down; their efforts were not rapturously received. The evening wore on and the *usquaebach* and vodka flowed until a member of the audience stood up to speak. Without an interpreter he spoke in English. From the outset it was clear that this was not about Lot or the airline business, it was about the freedom of flying—or more specifically, about freedom: the stretching of wings without game-keepers and shotguns. The picture was painted in vivid colours and the be-medalled Russian general sat there glumly and took it. The applause was thunderous.

*   *   *   *   *

Our American colleagues were being strictly rationed on fuel and it was affecting their customer relationships. With our Iranian fuel-oil business in full swing we were seeing more aviation fuel than we actually needed, particularly if Agip was going to help us to get Rome Airport going, so we devised a fuel operation whereby airlines could pick up supplementary fuel in the Caribbean without breaching their domestic restrictions.

Since one of our Japanese customers was also in Houston when we tied up the deal, we prevailed upon Bill Little, a magnificent American with an even more magnificent statuesque blonde wife, to organize a dinner party. Bill dragooned his hospitable boss to host it at his home. He was more than willing, especially as he had just bought a beautiful new sitting-room suite. The evening was hugely enjoyable; the Japanese joined in and the whisky flowed. Tired from his journey and soothed by John Barleycorn, the most senior of our clients fell asleep comfortably on the new settee. Until, that is, the hostess grabbed my arm gasping, 'Jesus!' and pointing at a swelling plume of blue smoke emanating from the gentleman's posterior.

Once the burning trousers and their owner had been comprehensively doused and we had applied many damp cloths to the furniture, the smoke eventually cleared. With apologies all round we retreated as our man stared mournfully at his new settee.

Aviation conferences, organized by Terry O'Keefe, offered excellent opportunities to drive home the safety and product-quality messages. That year in Puerto Rico Terry expressed concern about the South American delegates—not about their competence or their attention, but about their social participation. They seemed quiet and almost uninterested. After a corporate supper filled with the usual bonhomie, the delegates streamed off to bed, Terry in the lead. I decided to have another drink and a stroll on this beautiful warm evening. I noticed there was a lively party going on in one of the halls; great dance music was filling the room. Uninvited but unchallenged, I wandered into what was a party of office stationery executives who must have been celebrating a fabulous year. I pulled up a chair to watch, and to my amazement spotted the two South Americans—

unskilled in the sale of stationery, but enormously skilled in the tango and other even more exotic dances. Time and again they swept past me, enveloped in both the dance and their partners!

The next day Terry again sought my view about the seemingly apathetic South American contingent. They're fine, I said. They have no difficulty technically or commercially and they participate well. He agreed. Socially they live in a different world which we have no knowledge of. For all we know they could be passionate tangoists or masters of the paso doble. He nodded sadly—little did he know!

Back to Rome next for our weekend with Agip. Arnaldo, our travel guide, drove and his wife Norma was navigatrice. Having said farewell to Ruby Bambino the dog, she set the compass north to lovely Sienna. It was dusk by the time we made our way down the country road, through the vines, towards their vineyard home and a haven of light and warmth. Our hosts filled the glasses with their own vintage, and roasted partridges over a log fire. After several courses we reclined on soft settees around the hearth sampling superb cheeses. We were comfortable and beginning to feel drowsy when—bang! The door flew open and half a dozen young women burst in, followed by equally vivacious young men, all chanting 'We give you the corks, now where is the wine?' These were local medical students and their uninhibited energy immediately dispersed all lethargy. The hours passed merrily until dawn, when Jo and I enjoyed a memorable view of the sun rising over vineyards and acacias.

Back in Portugal, we had a full day. After receiving the performance, market, safety and quality reports we were ready for a bit of relaxation. Arnaldo and Norma, Bill and Ann came with us to dinner. Ann was in great fettle, asking where the excitement was that night. 'Nobody's on fire here,' I said. Then Arnaldo whispered, 'Sitting over there, Ann, is the King of Italy, Umberto II.' Ann was flabbergasted. A real king in the same restaurant? What a story for the folks back home! The king was with someone, and an aide was just leaving. Without hesitation Arnaldo acquainted himself with the aide, and before the next course was served Ann and Norma were being presented. The king was obviously delighted that subjects of his briefly held kingdom not only recognized him but were sufficiently impressed to wish to meet him. The women were invited for coffee the next day. Ann almost had to be carried back to her seat.

I began to wonder if this was an Arnaldo set-up. But no, there was a King Umberto in exile in Portugal and yes, this was his favourite restaurant. What was Ann to wear tomorrow? She had not brought enough clothes.

Why had not Bill told her she would be meeting royalty? And what gift could she take? Before long pictures of Ann Boleyn and Catherine Parr were being offered up with the advice 'Don't lose your head!' Norma would go as interpreter and chaperone. But the present? Bill's Shell Oil lighter was rejected as was the Shell-monogrammed hand towel and the Shell-embossed note clip. Arnaldo intervened: flowers were the answer. And so the day came and, bedecked with hats, armed with flowers and driven by a uniformed driver Arnaldo had organized with the hotel, they set off. Coffee was a success, addresses were exchanged—and Bill faced the prospect of a future royal visit. 'Does he smoke?' he asked, recalling his last international visitor.

A few years later we turned up in Houston and there over Ann's mantelpiece was a framed Christmas card signed by Umberto, and further over in the display cabinet, two more. If only George III had concentrated a bit more on the ladies, we might have kept the colony.

*   *   *   *   *

While the aviation market had enhanced its returns, the gasoline market, which I was advising, had not. It seemed lacklustre, and the advent of unleaded petrol, which should have offered some sort of promotion opportunity, had provided no increase in either volume or margin. We had to find a way to boost the business.

Our researchers were confident they had discovered an ingredient that would enhance engine performance. The tests had gone well so we had something to promote and sell, but we needed a platform. The oil crisis had cast a negative gloom over motor racing and everybody in our industry was out ... I hoped the discovery might offer a golden opportunity to re-enter the Grand Prix business. Persuading my German and Dutch colleagues was not a problem, and the UK was also happy to come along.

Our discussions with McLaren went well. Lauda was their No. 1 driver, and his car was No. 1. We would have a number one on the car's front and side, both next to our logo. The photographic opportunities would be fantastic—but we still had to persuade the board. This was a gamble we might lose, but then, having the No. 1 for a season and a young man called Schumacher as the No. 2 driver, we surely had the dream team! The board bought it and 1.6 billion people watched the Shell logo flash around the track. Shell stations carried Formula Shell. The market rose exponentially.

Prost took over from Lauda, and Senna joined them. Seven years out of eight they took the championship. Ron Dennis was a superb manager and

Prost was a magic driver with a lovely personality. Senna was a winner, a dedicated perfectionist. His death was a terrible blow. These drivers live on the edge and there is no better example of this than Nigel Mansell.

I was invited to present the winning trophy at the British Grand Prix. Mansell was the crowd's favourite but his car had had trouble, so he started in the second car, which had no air-conditioning. For a man of Mansell's girth, this was serious. But for my people there was a different problem. The trophy was solid crystal: heavy and valuable. We practised the handover, me holding tight with my stump and not letting go. I did not know until later that the practice was carried out with a prototype, not the real trophy—they didn't even trust me in the office! They were waking up at night in cold sweats seeing this costly prize slipping off the end of my stump.

And so there I was on the podium, memories flooding back of trying to put the winner's belt around Isaac Ikhouria's sweat-slippery waist, when a seriously dehydrated Nigel Mansell staggered up the steps to receive the trophy. I shook his hand and lifted the cup for the crowd to see before passing it to him. He dropped it—but after all my training and rehearsals I managed to stop it crashing to the floor! Following the presentation, Mansell went to hospital and was put on a drip, and I went for a cup of tea before driving home. There is no substitute for practice, whether it be driving, golfing, public speaking or handing over crystal trophies.

*　　*　　*　　*　　*

My two years in the centre had passed in a flash. The Iranian deal was still running and its positive contribution secure, as was the Latsis deal. The Grand Prix venture had been a success, not just in sales but also in its impact on staff morale.

Still, there were other businesses to run, so we moved on after a hectic, exciting period I would never have missed. Australia, which in time became a second home for the wider Reid family, was to be the next stop.

7

Australia

When you look back at the places you have been, certain pictures always come to mind. For Borneo it was Lubok Antu, the 'pool of spirits', and the morning mist rolling back to reveal a never-ending forest. For Nigeria it was the noisy, brightly coloured crush of bustling humanity at Murtala Muhammed airport in Ikeja. For Thailand it was the temple on the mountain overlooking Royal Hua Hin golf course, the king's summer palace and the sea. As we landed at Melbourne Airport, I wondered what would it be for Australia.

We arrived on a Friday and soon understood that the first imperative was to find tickets. The Melbourne Cricket Ground (MCG) was just in front of the Hilton Hotel where we were staying, but the much-coveted tickets were not for cricket, they were for the final of the Aussie Rules football: Richmond vs Carlton. The stadium the next day was packed, every one of the 74,000 seats taken. After dazzling pyrotechnics, the players burst onto the ground through their supporting banners and the game was on. As a spectacle it's fantastic: the ball goes from end to end in seconds and it is the only game in the world where you win a point for a near-miss. The photographs show huge men rising in the air to take incredible catches, after which they would boot the ball miles down the field.

The games are played on cricket ovals, which are different sizes, the MCG being the biggest. Fitzroy, one of its neighbours, is tiny by comparison, giving the nimble an advantage over the mighty. But that day in the MCG against the odds it was a 'tiny' man who foxed them all, pirouetting his way through the giants to poke the ball through the posts time and again. What a memorable introduction to Australia, Land of Sport!

* * * * *

Our head office was in Melbourne but my remit for the down-stream, the oil and gas production companies, took in the whole country plus Papua New Guinea, Fiji, Western Samoa and American Samoa. We had two refineries: one in Geelong just outside Melbourne, and one in Sydney, in Parramatta.

Australia's primary cities lie along the coast: Brisbane in Queensland, Sydney in New South Wales, Melbourne in Victoria, Adelaide in South Australia and Perth in West Australia. Each state had its own premier, parliament and elections. The federal government was based in Canberra, and elections for the national government were separate from those for the states, so a mixture of parties competed for federal and state government seats.

Our chairman, Leslie Frogatt, had built a successful down-stream business, some rapidly growing exploration and production activity and a metals business. Establishing a corporate reputation in a highly critical society is no easy job; doing it in the natural resource arena is a full dimension more difficult.

Frogatt was about to retire. His replacement, Kevan Gosper, was an outstanding young Australian, a former medal-winner in the 4 × 400 m relay at the Melbourne Olympics and a Commonwealth Games champion. Kevan's wife was Olympic swimmer Jilly, a truly delightful person we had known in London when Kevan worked there. Physically the pair were godlike, though neither showed any vanity. They were good people with the brightest future ahead of them.

Jo and Jilly spent one fun evening organizing a welcome-home dinner party. The following day, after the party, came the terrible news that Jilly had died in the night from a cerebral haemorrhage. This was a tragedy of Wagnerian proportions. Wotan had removed the fairest of his princesses, and all was dark. An event of this magnitude strikes deep into the heart of an organization. The impact on the new chairman was massive. Frogatt handled the situation with sympathy and speed, giving Kevan time and space to re-establish his life without the worries of business, and we all pitched in where we could.

*   *   *   *   *

Any activity in Australia, be it sport, politics or business, had to have a contest and an arena. The public expected it, loved it and—given half a chance—joined in themselves. The gasoline market was one such activity, the arena was the forecourt and the contestants were the press, the association of dealers, and the politicians, urged on by the public. If the prices were the same or close, the public yelled 'Collusion!' If they were different, the dealers clamoured for more discounts; and if they kept changing, the politicians shouted for stability. It was great fun.

New South Wales under long-time Premier Neville Wran opted for price control. Supplies began to run short during a refinery shutdown because the export price from Singapore (the price at which we were buying) was higher than the controlled price (at which we had to sell), so we did not ship. Faced with this hard fact Neville moved quickly and hiked up the control price; then there were no shortages. Later he admitted that price control was the worst decision he had ever made. If he put the price up he was vilified by the press; if he lowered it, it was never by enough, and always too late.

The petrol price saga went on and on and took me into the offices of premiers and the federal finance minister. Joh Bjelke-Petersen was the leader of Queensland's National Party. A former farmer from the outback, he had stumped his own land (removed the trees to create pasture: a tough job) and built his farm before becoming state premier.

Joh had encouraged Japanese immigrants by providing retirement homes and creating an attractive environment for the parents of wealthy Japanese executives. As an incentive for fellow Australians to invest and build homes in Queensland, he had removed inheritance tax, which was a state, not a federal, tax. This created a second-home resettlement rush that benefited the state economy and treasury, but did not endear him to either his fellow premiers or the federal treasurer.

It was against this background that I went to see him. I asked my Queensland colleagues how I should handle the interview, and they told me, 'Don't worry; you don't need to talk, just listen.' Luckily Bjelke-Petersen held Shell in high regard. Many years earlier we had given him an aviation fuel tank for his small plane, and that had kept him in the air. We talked about the international oil scene, OPEC prices and the Middle East, which caught his imagination. He believed in markets, as well he might with his growing metals and coal exports.

On our third quarterly meeting he was incensed about the federal government and declared secession was the only way. When he had finished his tirade I told him that one of the things I did know about was secession. All of a sudden he was a listener. In a few minutes I took him through the tragedy of Biafra, I then described in some detail Colonel Ojukwu's declaration, which had sounded much like his own, and the irresistible pressure he had created for himself by stating the secession option publicly. I pointed out that once you state you are on the way to it, you are liable to be driven by an avalanche of public pressure, and at that stage drawing back is political suicide. Ojukwu, I explained, had had no choice once the genie was out of the bottle. Bjelke-Petersen quietly thanked me and we said goodbye.

The premier of South Australia, a former eye surgeon, was concerned about the variation in prices. When he travelled to Victoria and saw the cross-state differences, he became even more concerned. For a man used to making micro-millimetre adjustments, this was new territory. He asked me to call an industry meeting to seek uniformity. I explained that at least two of our competitors would lose their pension rights if they were involved in any such discussion, and that in any case antitrust legislation carried criminal penalties, so it was a non-starter. He would have to accept that this was a market. We went over Neville Wran's New South Wales experience and, now calmer, the premier invited us to the chamber, which was in session. At the end, the local representative in true Aussie fashion said, 'You know, Premier, when I went horizontal and you fixed my eye you were great, but now I'm vertical and you're in a new job, I'm not so sure!'

The West Australia meeting with Premier Charles Court (father-in-law of the great tennis player Margaret Court) had a wider agenda, with an important North West Shelf gas project in prime place. Frogatt described the project's status and the support it needed. It was a good, positive meeting and West Australia certainly welcomed the investment. I talked generally about the international oil and gas scene and the situation in Iran and the Middle East—key markets for West Australia livestock.

Over dinner Charles told us about his experience as an officer during the War, and his role taking care of POWs after the Japanese surrender. Each morning he had led them in physical exercises and gymnastics. A few months ago, he told us, he had been invited to a Remembrance service in Tokyo: the first time in more than 30 years that an Australian had been included in their private ceremony. After the tributes had been made, the attendees congregated in a large room, and then out trotted a group of elderly men in white shorts and singlets to perform their physical exercises. The premier had been deeply touched by this gesture, particularly as a number of the men described how important the daily routine had been to their morale and eventual rehabilitation.

As we left, Court commiserated with Frogatt about the plight of the English in the current Test Match. Little did he imagine that Frogatt would sit up all night watching Botham score more than 160 runs to win the test match practically single-handed.

The Australians' passionate love of sport was coupled with an intensity at work that was helping to build the company assets and cement its profits. Petrol price wars and associated political concerns had stilled somewhat, and this part of the business was strong, had good leadership and a sound commercial strategy. We had reduced costs while retaining market share, so had enhanced our margins. Our new buildings and refurbishments were underpinning this policy as well as adding to the bottom line. We had also introduced a more sophisticated way of analysing the retail market potential, which meant we could site our new investments much more effectively.

Our biggest business risk lay in industrial relations. Union militancy, or failure to reach agreement on some point, had completely scuppered some activities. A Land Rover abandoned at the terminal in Queensland bore testimony to this: a dispute about driving had become so intractable that the perfectly good vehicle remained marooned where it stood, paralysed by intransigence.

The industry itself did not help. Under 'hospitality agreements', road tankers denied product in their home depots should have been able to load in a competitor's terminal. But the companies that were not on strike would frustrate loading in order to take short-term commercial advantage. This lack of cooperation needed to be addressed. The unions did not want a total strike because if the state ran dry they would have to face public anger and pressure from within their own ranks. However, after a considered programme of hard talking, and having shown our willingness to stand by our hospitality agreements, we changed the atmosphere and in doing so neutralized an unfair weapon.

One day the manager in the Clyde refinery in Sydney rang me to say he had caught an operative stealing gasoline. We agreed he should be fired immediately. We would not pursue criminal charges since this would only delay matters and anyway the judicial decision could not be guaranteed. It was done and dusted in 20 minutes. A couple of hours later the personnel director rang, making a case to take the matter to the commission—the industrial relations government agency that seemed to be involved in every dispute. In this process everything came out as a draw and appeasement was the medicine, which was all new—and completely unacceptable—to me. In a case as straightforward as theft, witnessed by the boss, the path we had chosen was the only one open to us; we stuck with our decision. The professionals were not happy—a draw meant you could not lose. Very un-Australian you would have thought, but the question of who owns Australia, the workers or the bosses, runs deep in the national psyche.

The refinery's major maintenance shutdowns were not normally too difficult to manage, though they could be punctuated by official and unofficial stoppages. The start-ups were where the real battles lay. Fortunately they happened only once every three years, but on occasion the company had been held to ransom and had to pay for the spade to be removed from the crude oil feed line before units could get back into operation. This time we had had two stoppages in two weeks. As the crisis began to build, I held a review. On the relatively small site concerned we had contract labour of 750. The next stoppage was about imported labour: 'scab' workers. A sub-contractor was using 20 or 30 workers from the Pacific Islands, so the other men walked out. It quickly transpired that the 'Islanders' were in fact Australian, and far from being scabs they were long-time paid-up members of the union. This embarrassing press revelation saw the men return to work, but the union then had another go, declaring that the extra men were paid not as individuals but as individual corporations. For us this was the final straw. We told the contractors that we would finish the job not with 750 men but with 350, and that they should prepare for that. By now it was ten days before Christmas, so we would start in the New Year. Our tanks were full of imported product at workable prices, so operations continued.

The unions then came back asking for work to be resumed and claiming a Christmas bonus in spite of three weeks' disrupted work. They were serious—but so were we. As the temperature rose, the press revealed that if the individual corporate ruling were to be conceded to the unions and remuneration paid to the individual rather than the personal company, then most of their members would be worse off, since they used the corporate device to claim for their cars, fishing boats and so on. By now the men themselves were at war with the union. This debate strangely allowed the work to resume without disruption, the job to be finished and all the units to restart without any financial ransom being demanded or given!

This episode was a watershed, and from then on management began to manage without constantly looking over its shoulder for unwanted advice—although of course it did not mean that all the union troubles were over. We had ordered a product tanker specifically for the Clyde refinery because the approach through the creeks was difficult. It was ready in Mitsubishi's yard in Japan. The crew set off to collect it—via Hong Kong to break the journey(!). When they were about to take over the vessel they learned that two of the Mitsubishi engineers were to travel with them on board to Australia. The union demanded that in that case we must hire an additional cook. The boat then sat in dock until the issue was negotiated out, the union apparently unembarrassed that a sophisticated, state-of-the-art vessel, delivered on budget and on time, sat there doing nothing and going nowhere. At last an agreement was hammered out, facilitated more by a wave of homesickness than any concessions.

The postscript was happier than the story. When the boat arrived back at Clyde and loaded, the rest of the industry's coastal tankers were paralysed by a nationwide strike. Our vessel, having already had its strike, sailed on and took advantage by filling all the emptying tanks around the coast.

We were not alone in these troubles. In the middle of a long-running strike, one of our competitors was suddenly required by the state government to hand over the offending workers' personal details. Armed with the strikers' names and addresses, the state police knocked on their doors at 0600 hours and handed them an ultimatum. As the men spluttered, Mum put the breakfast on the table and pointed to the door: 'Back to work!' was the order. This was a robust society!

These high jinks did not deter our investment programme. The North West Shelf gas project was ready to go and in the absence of the key executive I had the pleasure of getting the board's approval for it. Our coal and metals businesses were growing fast; we were well diversified and progressing projects across a wide front. It was an exciting business in an explosive phase of growth. Subsequently as we divested our metals and coal businesses the scope narrowed, but the slack was taken up by gas. Australia could have become a mega-conglomerate on its own, but the possibility was lost in the corporate focus on oil and gas. This was what the market wanted, too, so although the management in Australia would have made an outstanding success of a conglomerate, that avenue was closed.

*   *   *   *   *

It was already very hot in Melbourne when the fires exploded. The forest was a tinder box: lightning struck the match and the bark covering the ground provided the kindling. A searing wind surged through the forest and sap-filled gum trees exploded, each carrying the fire hundreds of yards forward. A young team of firefighters was caught and killed in the accelerating inferno. The elemental power of nature dominates this continent, which is prone in all seasons to devastating fires and floods. Australians have a certain steel in their character and a respect for their environment, which they both love and fear.

My responsibility covered the Pacific Islands, where they were pleased with our service and paid most of the time. Tonga, though often whipped by typhoons, seemed unfailingly good-humoured—no evening there was complete without a session around the piano singing the hymns and psalms of our youth. The remarkable timbre of Tongan voices made ours seem weak by comparison, but Jo's singing experience helped us share involvement and participation across the continents.

Fiji was a much bigger market and more competitive; its tourism business and economy were growing fast. The island's population had been supplemented by Indian immigrants, whose numbers were split in religious terms between Muslim and Hindu. Their combined numbers put them in the majority, but they did not vote together until later, at which point the original population was not best pleased to find itself in the minority!

Fiji's finance minister, a former rugby international, dearly wanted a renaissance of Fijian rugby, and his wish was granted in the seven-aside game. His knowledge of international affairs and grasp of the oil situation was impressive: with men of this calibre in control, the country was in good hands.

Away above Queensland across the Coral Sea is Papua New Guinea (PNG), a land mass crumpled millions of years ago by tectonic forces. The concertinaed slopes are clothed in jungle. The people live and plant their crops in the valleys between, each group communicating in a dialect that is incomprehensible to the residents of the next furrow. In these fractured communities there has been a culture of reciprocal violence. Physical harm, no matter what the cause, intentional or unintentional, calls for payback: tooth for a tooth, a head for a head, in what can be an incredibly difficult sequence to halt.

PNG's first Sisters of Mercy came to Goroka from Australia in 1956. One elderly sister, talking of her work and the many years she had lived and taught in her adopted country, described how things went wrong. There was always some 'buggerup', she said, which expressive term has now entered the pidgin English lexicon.

PNG is rich in mineral wealth, which is slowly being discovered and mined. The jungles hold the secrets—just as they conceal the story of a past war in which many Australians and Japanese died in horrific, ugly battles. But the former enemies now need each other to fuel their industries and equip their homes, their roads and their factories, so they 'live in peace and not in pieces', as our philosophical Nigerian gardener used to say.

PNG's small operation was run by an effective team of nationals and Australians who understood the country and its peoples. Port Moresby hosted the Australian cricket team every year and its own team had a creditable record, winning some outstanding victories on pitches that favoured local knowledge.

*   *   *   *   *

When the time came for us to leave Australia, we left one son settled with his family and another heading the same way. Twenty-five years on they are Australians, as are the second and third generations of antipodean Reids. It had been an unforgettable experience during a period of explosively fruitful development for the company, living among the resilient people of that huge, beautiful continent.

8

Shell UK

It was a bright afternoon and everything looked to be in good order when the phone rang—the chairman Pete Baxendell wanted to see me. All sorts of ideas ran through my head. Perhaps he wanted to discuss a new beginning in Iran? Or a fresh initiative in Australia? Or better still a Chinese project, which could open up all sorts of doors? It must be serious, otherwise he would have told me over the phone.

In his office he came quickly to the point: 'We want you to run Shell UK.' Pete had previously done this job himself and was familiar with the problems. He was a professional up-stream man and he knew the North Sea operation well, right from its beginnings. I knew the down-stream because I had advised on it. We discussed the proposition briefly and the timings.

Just before I left he described a problem he was handing over with the job. 'Sullom Voe, our export terminal, is built on Shetland land,' he explained, 'but when we acquired the land we didn't. The money was paid and accepted, and it was a generous sum for what had been sheep-grazing land. But now that it's covered in tanks and pipelines and has something more valuable than sheep on it, the landlord says it's time to make a proper agreement. What's more, the business has become intensely political, which puts the price up again. Over to you!' I thanked him for that—and for having the confidence to let me run Shell's second-biggest company, Shell US being the biggest.

My predecessor, John Raisman, had put together a very tidy and helpful governance structure with three experienced board members: Sir Tom Risk was governor of the Bank of Scotland, a lawyer by training, and still active in the development and financing of the North Sea operation; Sir Francis Tombs, an engineer with a wealth of experience in the electricity industry and chairmanship of Rolls Royce, and Fred Holliday, vice-chancellor of Durham University. We had a strong team, and my first job was to get to know them.

The down-stream was our primary engagement with the public. They filled up their cars at our stations throughout Britain. Each time they did so they resented how much it cost them in money and in time. Self-service was not a positive but a negative innovation and if, because of industrial action, they were denied the doubtful pleasure of filling up at all, all hell broke loose and reactions were influenced and recorded by the media. Having over 20 per cent of the market, Shell naturally shouldered over 20 per cent of the problem.

The refineries Shell Haven and Stanlow were substantially overstaffed and their unions were resistant to change. This would be a battlefield, but we could not ignore it. The up-stream was a technical achievement of the highest order run by experienced people, and its cash flow reflected that. Questions remained, however; could the investments—particularly the new and inevitably more expensive projects—meet the criteria in an uncertain price regime? Were there cheaper and more imaginative ways of accessing the new fields?

One way to offset the cost burden was to choose the right pricing formula for the gas. The government take offered another opportunity: more activity could earn us more favourable tax measures. One great advantage this business had was its partnership with Esso, which not only had financial strength, but also understood its non-operator role. They were interested in and knowledgeable about what we were doing, and their support came with a challenge, which was good for us.

The chemicals manufacturing at Carrington was working in a graveyard of equipment well past its sell-by date, and this also needed to be dealt with. John Collins, my chemical manager from Nigeria, was in place and the right man to do it.

I went to bed that first night with many thoughts, but uppermost I knew we needed to communicate and share our problems with all our colleagues and solve them together. Face-to-face meetings with employees were essential if we wanted to engage them. One important factor in our favour was that Britain was changing. 'The future is in our hands' was a clear message running through the country.

First we needed to sort the problem of sheep or non-sheep on Sullom Voe. Court cases were being prepared, the local press was hostile and it looked as if we were in a downward spiral. I needed to stop legal activity and suspend press and public relations. Sullom Voe was a worldwide name in the industry but it was built on a fig leaf, so far unpriced. Though we were one of the biggest and best organized corporations in the world, we had failed to negotiate and sign a lease for the land on which we would make a huge investment and construct the key installation for exporting North Sea oil.

I needed to introduce myself to the Shetland Islands' council convenor. A slight man with lively eyes, he swam in the sea every day of the year and you could see he was very fit. He spoke quietly with careful articulation. I had taken care to learn about him—but not from my troops. Their views were coloured by the lack of progress to date and by their skirmishes with the islands' chief executive who, despite his posturing, had no authority to make a deal. My uncle had been postmaster in Lerwick for a number of years and had loved the place. A gregarious man, he had made friends easily and, serendipitously, had been acquainted with the convenor.

'What you have to understand, Bob,' he said, 'is that he has the interest of the Isles at heart. To him it is a mission, not an office or a job. He is determined to protect the assets and sees that as an obligation to future generations. He has a place in history and he plans to fulfil it. If you want to begin a conversation, start from there, and listen.'

When the convenor and I met at last, I had made up my mind to say nothing about Sullom Voe but to engage him on the island and its history, rather than its future. He seemed interested in where I had been educated, why I had joined Shell and what I had learned from the variety of countries I had lived in. I talked first about Nigeria, about 'speaking the bitterness' and then moving on. I told him about General Gowon and his actions at the end of the civil war that had led to reconciliation and reintegration. These were my hopes for South Africa when apartheid came to an end. Seemingly encouraged by these narratives, he talked about the Vikings and which way they should face in modern times. We have rich peoples on all sides, he said, and we should keep good relationships with all of them. He described Up Helly Aa, the Viking Fire Festival, which takes place in Lerwick every year. A boat built over 12 months by the community is set on fire on festival day and launched out to sea. It is a day, he said, for long and deep libations, and care needs to be taken. I said my uncle had warned me well. He then formally invited me and Jo to the festival, and we parted on good terms.

A few days later my team anxiously enquired what had we agreed to. Had he accepted their points in any way, and could we write a letter? To their amazement I said I had not discussed it. 'But you were there for over two hours! What did you talk about?'

'Scotland, Nigeria, a bit on Thailand, the Viking Up Helly Aa and his invitation.' They left, bemused, shaking their heads.

And so to the festival—and what an evening that was! I met the convenor in the early part of the day and we toasted the boat and the celebration. The vessel was a magnificent piece of work and I am sure would have survived a long Viking expedition. The braziers around it set a ghostly background as the sails billowed in the stiff breeze. The noise of the wind, the revellers and the proximity of the sea made conversation impossible, but we watched the boat on its blazing way until it became a distant glimmering speck. We then retired with the boat builders and their supporters to the school hall, where mountains of sandwiches and urns of soup were waiting. We left for London the following day at dawn.

I wrote to the convenor to express our enjoyment of the occasion, thank him for the privilege of being there and to invite him to dinner in London. He accepted and a date was set. The legal team put the final touches to our offer, emphasizing our desire for an early settlement since uncertainty was not helpful to either party. The offer was a good one, and so it should be, for we would be using that facility for a long time. The income flow was clearly assured to secure provision for future generations. It was hand-delivered to the convenor's officers.

Before the dinner I stressed to my team that the deal was not to be discussed. They could not understand this and did not like it, but those were the rules. The convenor arrived and after a few pleasantries we sat down to eat. We talked of the various countries where my team had worked—Brunei, Nigeria, Australia and the States, among others. As before, he was interested, and he engaged the interest of these hard-headed engineers and geologists. They in turn asked him about the Shetlands, its ambitions and its relationship with its neighbours.

As we came to the end of the dinner he took a small glass of brandy and, silently commanding the attention of the table, said 'There comes a time when the wind falls away and the waves are quiet, and that is the time when the fishermen makes his catch. And later the wind returns and the boats set sail together to make another harvest.'

I replied, 'Wisdom gained through the ages enables the Shetlanders to know "when", and it is this we respect. Thank you for coming.' With that

Delphic exchange he thanked us for dinner and bade us goodnight. I saw him to his car and he left smiling.

When I returned upstairs my table companions were restless and antagonistic. I told them, 'Gentlemen, you have a deal. You may not see it now and later you must ask yourself why. But remember, while oil industry activities are in the hands of technocrats, its assets are in the hands of poets and politicians. They have spoken to those who can listen.' The offer was formally accepted and the deal was done. The only losers were the lawyers, since no more fees were needed.

*　　*　　*　　*　　*

I had not long been in the job when board member and Bank of Scotland Governor Tom Risk proposed to the chairman that I would benefit from banking-industry experience. Tom was persuasive, and before long I became a director of Bank of Scotland. He was also right: it was a useful learning and formative experience, and its value went on far beyond my career at Shell.

*　　*　　*　　*　　*

Stanlow was a major refinery that sat on a large piece of land close to the River Mersey. It was a tidy, well run site; its record was good and its plant performance excellent. But in a fast-moving, competitive world its costs were too high; it had too many people, hence productivity was too low. I needed to make the management realise this, then the executives and the workforce.

Each year the company performance was presented to the staff. It was a great story about the North Sea developments, the down-stream refineries and the petrol stations, and then a description of what we were doing on community relations, education and the arts.

In my first year I changed the focus of the down-stream segment. We concentrated on competitive positioning—our plant performance measured up well, our safety record was good but our productivity was at the bottom of the league. One thing the man on the ground understood well was leagues. Liverpool or Everton, wherever they were in the league, they had to do better if they wanted to reach the top. We were at the bottom and something had better be done about it. So the scene was set and review after review worked through programmes to reduce manpower. Some redundancy was inevitable, but normal wastage and moratoriums on recruitment also helped. In a few years our position in the league table had significantly improved.

Down the road in Carrington, John Collins was doing the same. His graveyard of dead plant was cleared out, giving the site more space and a new look. It was on conclusion of this work that something special happened. Jon M. Huntsman, a manufacturer and marketer of speciality chemicals, had many years before gone to McDonald's headquarters with a locked briefcase chained to his wrist. Once in the boardroom, he had unchained and unlocked the briefcase and placed in front of the board the prototype 'clamshell' container whose successors would ultimately hold millions if not billions of burgers. His proposal was accepted unanimously. One can say the rest is history, but there was more, and we were part of it.

Jon turned up in London wanting to buy our polystyrene plant and a warehouse to use as his office. We sold him both, removing more loss-making plant from Chemical's bottom line. Further profit came not in cash but in example, just when we needed it, at Carrington and up the road at Stanlow.

The foreman in Jon's new warehouse packing area went off sick. Jon went downstairs, rolled up his sleeves and set to. By lunchtime his secretary was in dungarees and also working on the floor. In a world of demarcation where job-shares were impossible and tasks remained undone because shop stewards could not agree who should do them, this was incredible. This new attitude raised productivity and eliminated disputes. In that area of Merseyside news travels fast, and as the disputes disappeared in Carrington, so did they in Stanlow. It also inspired other new approaches, and one major one was in the area of redundancy.

Madeleine McGill, a personnel consultant working with John Collins, persuaded him to look at redundancy in a new way. The idea was to see it as an opportunity, to look at life anew and consider what you wanted to be and do. I bought into this right away and the programmes we put in place had some extremely positive results. Once people stopped looking backwards, they came up with a whole range of ideas outside the limits of their past employment. 'Take the fear out of redundancy' was the idea, and it worked. There was cynicism to begin with, but with each individual success it began to disappear. A well-known person in admin, for example, said she wanted to be a long-distance lorry driver, and after some weeks' training she proudly displayed her licence and set off on her new career up the M6!

*   *   *   *   *

Early one morning I got the message that a burst pipeline in Stanlow had leaked oil into the Mersey. I was there in a few hours. The management were in shock, and it took time to settle them down; the media were everywhere like terrorists. I need to know what had happened. First of all—which pipeline? It was the one that carried the 'Venezuelan heavy', the primary feedstock for bitumen. Second, where was the leak? Someone produced a plan that showed exactly where the break was. Third, can you confirm pumping has stopped? This question raised hackles. 'Of course!' was the answer. Fourth, when did you stop? There was some uncertainty here, which is when a picture began to emerge. The break had been worsened by the pumping. Venezuelan heavy was, as you would expect, heavy. It was normally heated to make it flow more easily. You could see that around bitumen plants where all the pipelines and connections are wrapped in heat-preserving material. Fifth, is the pipeline clear? If it's not, it will be very difficult to make it flow again.

A helicopter flew us over the ruptured pipeline. It was not the gaping hole one might have expected, and clean-up action was already under way. This was not a question of ladling it up; because of its weight it had to be dug up in chunks. We then flew over the Mersey, out to sea westwards and then up beyond the bridges eastward. There was no floating mess. The stuff had sunk.

It was time to take on the media. They were all at the local RSPB office, photographing the one bird with oil on its feet. The ornithologist in charge was incensed. 'Don't they realise flash photography will kill it? Its blood pressure is already through the roof!' I took this message and told them we would reconvene in ten minutes away from the centre. During the interview I stressed that so far only a single affected bird had been found, but that did not stop the reporters digging into their archives to produce pictures of any damaged bird, irrespective of whether it had ever seen the Mersey, far less flown across the Atlantic.

The topic then turned to spraying. I briefed them on the damage spraying could do: the chemical used would be toxic to fish. In this case anyway it was not relevant—there was nothing to spray. Disconsolately, they left. The facts were unexciting, still their imaginations were intact and they wrote their stories.

My last call was to the Wildlife Trust to confirm the leakage was stopped and to brief them on the bird beset by paparazzi. They accepted this with thanks and then asked the question that was really worrying them. Should we defer the start of the wildfowl shooting season? Here was I, concerned about the blood pressure of one poor bird which I would have taken home to look after had it not been in better hands, and there was the Trust concerned that their annual massacre might not start on time! I bit my tongue and said that in my opinion there was no need for delay.

It had been a long day. Peter Holmes, a main board director had come with me for, as he said, his education. He told me later that he had been reprimanded by the chairman for having been there. 'As a main board director you have no corporate responsibility and no role in the incident—that responsibility rests with the management and their board.'

The logic of this was brought out in two subsequent events: the *Exxon Valdes* oil spill, in which the chairman avoided direct involvement so did not confuse the lines of responsibility, and BP's Deepwater Horizon spill in the Gulf of Mexico, where the chairman became involved and confusion reigned for many years. Clarity is key in high-tension situations and roles need to be carefully defined. To be fair, Peter got his education and took no part in the media discussions.

*   *   *   *   *

Over the years we had been involved with the British Academy of Film and Television Arts (BAFTA), primarily in the production of industrial training films, which had helped modernize our work practices and improve safety performance. BAFTA's finances were never strong, despite the quality of their work, and a crisis had arisen. Its chairman wished to come and see me.

Sir Richard Attenborough duly appeared, open and cheerful, immediately recognizable from all the great films he had starred in. His predicament was clear but it was not all about money; it was also about competence and control. BAFTA had style, and style is expensive. It needed to change if it was to survive, and with better management it could. We agreed to help financially and to put somebody in to look at the costs and the effective management of their assets, which were not inconsiderable—particularly the historic building sitting under-used and un-merchandised in the heart of Piccadilly.

I chose Stephanie Chew, a highly able Oxford graduate with a keen commercial mind and obstinacy of purpose. Both were needed, as we soon found out. She and I made a point of attending BAFTA's events to get to know the set-up and try to identify where their style was outflanking their budget. One evening we went to an event at Hampton Court held in honour of Olivia de Havilland. The Princess Royal was to attend, as were all the stars of today and yesterday—a galaxy of talent if ever there was one. As we arrived we were enveloped in a swirling mist, arising, we assumed, from the river, but appearing nowhere else in the vicinity. As if the venue weren't dramatic (and expensive) enough, BAFTA had also hired an industrial-sized dry-ice machine. The princess pointedly remarked, 'I see the cost-control measures are beginning to bite.' We assured her they soon would.

Lesson number one was there: it was important that the academy's celebrated panache was converted into revenue. When next they set off to Cannes for the film festival, we transformed the journey into a vintage car rally through France, visiting Shell stations on the way where merchandising and filming opportunities had been set up. What would have been a cost exercise became a revenue-winner. And so it went on. BAFTA's building in Piccadilly became a prestigious venue-for-hire and began to earn handsomely, and the budget discipline Stephanie imposed started to take effect. Soon we were well on the way to an even keel.

South African apartheid dominated the newspapers at that time. In Shell we were determined to stick it out. We had made our views clear to President P. W. Botha and he had responded clearly: he did not like them. But our resilience was a positive even if he and his colleagues found our position unacceptable. Sir Richard had just made the film *Cry Freedom*, which predicted a better future with a charismatic leader who would preach reconciliation. This, of course, Nelson Mandela did, in a wonderful way. The ability to forgive is more evident in Africa than it is in parts of war-scarred Europe. The world saw it clearly with Mandela and earlier with General Gowon after the bitter civil war in Nigeria. Talking to Sir Richard about *Cry Freedom* and Mandela, I sketched out our problem in South Africa and our determination to stay. I told him of the letter we had received from Botha, which had been blunt and devoid of the courtesies one might expect of correspondence between the head of state and one of its biggest and oldest investors. Richard said he would talk to Mandela. He did, and the message came back that we were welcome to stay: Mandela said we could be part of the new South Africa and this would be made explicit to his political supporters. And so we did stay. One good turn, it would appear, deserves another.

The film industry is a tremendous business. Movies transport you into another world and into other people's lives. They create laughter and tears. You identify with people and their problems, their successes and the strains and stresses of their time. Watching *Lincoln* gives you a year's history lesson in two hours. *Mrs. Miniver* illustrates the futility of war, and *Amadeus* provides a glimpse of a genius. For an hour or two you share other people's sadnesses, their triumphs, their loves and their lives. It gives new perspectives to your own.

The great advantage of involvement with extra-curricular activities is not just the break it provides from the day-to-day problems of your enterprise, but also the different kinds of dedication and expertise you find in other organizations which you might want to build into your own. This was the case with BAFTA: it was an education.

*   *   *   *   *

The refineries' 'matching resources to needs' programmes were by the mid-'80s under way, if not yet fully accepted. The team leaders were all committed. The face of the company however is its petrol stations, and here the selection of the dealer is all-important. Just as we had established in Nigeria, retail is all about image: the forecourt should be clean and uncluttered, the pumps must all work—the shop must provide what the motorist wants and settlement must be quick.

Sir Thomas Farmer, who set up Kwik Fit, understood this well and based his business on these principles. Since he and Shell had a couple of competitive businesses in the car-servicing sector, I went to meet him to see if we could do something together. This was enlightening. He described how his selection of service-centre staff was based on in-depth appraisals of their achievements. He quoted an example of how one of his centres was in decline. 'I moved the old manager to an area that was stable but had little chance of growth,' he said, 'and put a put a new, top-flight man in the old place. The moved manager held the stable centre well, and the new man in the ailing centre quickly achieved substantial growth.' Farmer believed that there are £10,000-a-month staff and £50,000-a-month staff (these were nothing like his true figures—he was too canny to give that much away!). If you put the £10,000 person in the £50,000 spot the business will sink, but if you put the £50,000 person in the £10,000 business, it will flourish.

I thanked him for his advice and asked him if he was interested in acquiring our car-servicing businesses. If he was, I said, that was a question for another day. On my way back I understood why we were losing market share. Our supervision simply did not measure up, and it was not going to improve. It was time to look for a profitable exit.

We reviewed our position and our strategy in the petrol market. (Our executives called it gasoline but our customers called it petrol: that tells you something.) A group of us sat in on a presentation that highlighted why buying petrol was such an unattractive process. First, you had to break your journey, which meant delay. Secondly, removing fuel caps and handling dirty pump nozzles, often in doubtful weather conditions, was unpleasant. Then there was the faff of queuing to pay, possibly demands from within the car for sweets and other goodies displayed at the till, and finally you had to hand over your cash. We could identify with each piece of this unappealing process. 'And what do you get at the end of the whole unrewarding business?' The presenter asked. 'A receipt!'

This got us thinking and I called for ideas. In brainstorming sessions anything goes, but one thing was clear: reducing the price was not the answer. A price-cutting campaign would eat into the capital we had set aside to smarten up and improve our stations. We also realised we had come up with something—other than a receipt—that the customer could take away from the experience!

And so the idea of Shell glasses was born: a free tumbler with every fill-up over so many gallons. From day one these glasses were a winner— volumes soared. If little Johnny broke a glass, you could get another one the next time you filled up, making the exercise a minor opportunity rather than a major nightmare. At a party one evening the wife of the chairman of a big competitor took me aside and said, 'Please don't stop the glasses— they're the antidote to the destructive activities of my grandchildren!'

Idea number two was to extend this to the diesel market and to lorry drivers. Not glasses this time, but decanters in exchange for a fill-up. To my amazement this took off, and again had the most beneficial effect on our market share and our returns. Years later, after I had left Shell, I attended a lunch given by British tycoon Gerald Ronson. I sat opposite a man I had never met before. He said, 'I have always wanted to meet you!' I thought I felt warmth, but was mistaken. 'I own a chain of shops that sell household goods—kitchenware and glassware,' he went on. My uneasiness grew. 'You ruined my glassware business; I didn't sell a single tumbler for three years!' With that he returned to his lunch. A variety of replies formulated in my mind; I reflected on them but decided to talk to my neighbour instead. She simply said loudly, 'I loved them!' Before our companion could reply, Gerald rose and proposed the loyal toast, opening the way for a postprandial eulogy by one of his guests, after which the frustrated tumbler-magnate and I stiffly nodded goodbye.

Shell UK had multiple strengths in—among other things—technology, geosciences, engineering and operational research. I was determined that the corporation should be bigger than the sum of its parts, and created a small team of highly skilled people to work out how we could achieve that. First, we concentrated on the retail market.

Moving on from the successful sales initiative, the team came up with the concept of securing customers by facilitating their grocery shopping. The idea was to receive the order by phone and have it ready for collection. As the team enthusiastically laid out their plan, my mind went back to my father's butchers' shop, and how much went into putting customer orders together there. The range of items was minute compared with a modern grocery list, yet it had still been a time-consuming business. When the time came to review the team's work, I took them down to Junction 14 on the M4 where, close to the exit, Sainsbury's had a warehouse. I had arranged for them to see an order being prepared for a store by automatic recovery arms. They saw the space that would be required at the service station for the range of stock, the time it took to complete the order— and consequently the disruption and delays that would be involved. They realised the idea was impractical. Keen that they should not be disheartened, I suggested we look again to see if there were any parts of the project we could use.

They returned the following week to highlight one element of the grocery project that could have legs: the proposed antenna-based communication link that we had not got around to talking about at the first meeting. I was intrigued, and gave them the go-ahead to develop the idea. Before long they had worked out a network and devised less rudimentary instruments. They christened it 'Rabbit', and I could see its potential was enormous. For me it met the basic human need to talk, and to talk outside your office or your house without having to find a phone box. Also, unlike our current business, it had no danger of pollution. It would require significant investment, but that was no problem for us with the North Sea cash flow growing. It would certainly have wide public appeal. We pressed on.

In the meantime, I talked to my German colleague, Hans Pohl. He had been one of my supporters on the Formula One initiative and was enthused by the communication idea. He said Germany was developing similar ideas and consortia were being formed. We put a plan together and presented it to the directors. It got a good hearing and then the debate started. 'It's not your business' was one strident view. Eventually the key executive, with sincere apologies for not backing an idea that clearly had potential, declared his hand. 'I have no doubt the opportunity is there,' he said, 'and I appreciate your bringing it to us, but we have other opportunities in which we have established competencies and which absorb our cash flow. I regret we cannot pursue this one.'

So Rabbit became Vodaphone, and the rest is history. Did Shell make the right decision? I see the logic, but I have a feeling that its founder, Marcus Samuel, the original purveyor of shells from the Far East, would have taken it further, even if he had sold it at a higher price to make a huge return on his investment. But Hans and I had given it our best shot. It was fun to do it and it certainly gave us—in retrospect—something to dream on.

*   *   *   *   *

The exploration and production operation was run from Aberdeen and led by an energetic oil and gas engineer, Ian Henderson. I knew him well from the Nigerian oilfields where he built pipelines and tanks in the swamps of the Niger Delta. He was instrumental in achieving the massive increase of production during the first five years of the '60s, and his enthusiasm and drive had not diminished by the time he reached Aberdeen.

His responsibility for the platforms, production and pipeline facilities was great, but his responsibility for the people involved was greater. The platform managers, many of whom had seafaring backgrounds, were by personality and experience strong leaders. They commanded complicated static ships far off in the North Sea. Within their teams were drillers led by experienced tool-pushers, petroleum engineers who monitored and directed the drilling activity, and production engineers. These people had made the steel structure an economic growing concern, and they controlled the flow of the highly combustible liquid as well as the gases that arrived at the well head. The danger was not from the wild sea around them, but from what was coming from below the seabed.

Servicing platforms by supply boats was tricky, demanding first-class seamanship, calm waters, and above all, patience. Movement of people by helicopter was equally hazardous. There was constant communication between the platforms and the shore, and information was continuously updated. If an incident put the platform or staff at risk, a management team was immediately convened to concentrate solely on the event in question.

They provided the link to the outside world as well as coordinating assistance and action for the platform staff. This system, with no gaps in the line of command, had stood the test of time.

It was in action when a dumb tanker (a tanker with no motive power) that acted as the Auk field's storage facility broke free during a heavy storm and set off down the North Sea, its broken pipeline trailing behind it. These potentially disastrous incidents always seem to happen during a festive season. This time it was Christmas. Even with the pipeline acting as a semi-brake, the tanker made good speed towards the point at which we reckoned it would cross—or perhaps not cross—the sub-sea gas line from Norway. The line generated a substantial proportion of Britain's electricity. It may have been the strength of the storm or just luck, but fortunately it crossed the line without doing any damage, by which time the incident team had marshalled tugs to haul it off to Norway for repair.

A few years later a freighter in the Channel lost its steering and its power and started to drift towards our major gas platform. I sought action from a reluctant Department of Energy, urging them to instruct the crew to abandon ship and then, when all hands were off, to bomb and sink the vessel. As the department jittered, the boat moved off the collision course and away. Nevertheless I asked them to prepare an emergency procedure for any future incidents—but I never saw a draft of it.

Determined to improve his organization's safety performance, Ian launched the Target Zero initiative, in which every lost-time accident was logged and analysed. Platform managers were soon competing for the best record. Safety glasses became compulsory, and a new designer set was produced to forestall the excuse that they did not fit, or the lenses were too scratched. Wearing them became as automatic as wearing a safety helmet. Ian's achievement set an example to the rest of the Group.

In the midst of this our two platforms, Tern and Eider, were nearing completion. Their construction had been agreed with our partner Esso when crude prices were well below break-even point. The platforms had been built using modern project-management techniques very different from those involved in my first experience in Borneo. Managers were appointed; they assembled their teams and set up their offices well away from the main office. They were then on their own, seeking help from the centre only when they needed it. The process therefore had a beginning and an end and there were no distractions, no double duties or split allegiances, and it worked exceedingly well.

When Tern was finished we decided to ask the Queen Mother to open it. Our plan was to have the opening in an Aberdeen office, making the telecom and televised connection to the platform. I made a formal request to the palace outlining the day's programme, and received a speedy reply. The Queen Mother will be pleased to open the platform, it said, but she will do it properly: offshore. The risk of taking this venerable lady offshore, which would involve a long helicopter journey and then climbs up and down long flights of steel stairs and through narrow passage-ways, was too great. Enlisting some palace support I went back and strongly suggested that it would be better organized onshore. After a short delay, Her Majesty acquiesced.

When the day came the Queen Mother arrived, as always in excellent form, telling me with some humour that the provost had rung up to announce he would meet her at the city boundary. She had replied, 'I have been coming to Aberdeen since I was a little girl and nobody ever met me, and certainly not at the boundary, so there's no reason to start doing so now.' The helicopter trip was not mentioned and her speech was much appreciated by the crews and staff. As she left after lunch the colonel attending her thanked me warmly, saying, 'This certainly beats standing up to my waist in freezing water as we try to catch fish!' This was some woman!

*   *   *   *   *

Target Zero's progress showed in the statistics as well as in the standard of work. Another quality initiative, 'Right First Time' was producing a consistently high standard of workmanship. The idea was to stand back and plan, then remove the obvious risks before you begin. It sounds so basic, and yet it is so often ignored. As time went on I found myself observing these principles when facing household duties. Cutting the roses with safety glasses on may have amused the family, but getting in closer also did a much better job!

I was sitting in a meeting in Shell Centre when I got the message 'Helicopter down!' Within 40 minutes I was in a plane going north. The incident-management team briefed me. In the latter stages of its flight to Sumburgh on the southern tip of Shetland, a Chinook had dropped into the sea. Forty-seven people had died. There were only two survivors: a young engineering trainee and one of the pilots, an Indian who had survived the India–Pakistan war. Nobody had drowned; all the dead had been killed by the impact. The bodies were being brought ashore and, by nightfall, all but one had been found. Twenty-four hours later this last one emerged well over 100 miles away, having been carried on the powerful streams produced by the confluence of the Atlantic with the North Sea.

Unlike a fire, a crash is over in seconds. There is nothing to fight or to remedy, nothing that can be done—no more will happen. It is time then to look after the living, bury the dead, and seek the cause.

Aberdeen churches are bleak and formidable, clothed in their granite-grey walls and decorated with testimonies to those lost in past battles and tragedies at sea. The black-clad mourners filling the pews set the scene for an expression of sorrow for which there seemed no alleviation. As I waited for the service to begin my thoughts wandered back to that first helicopter crash in Borneo. Those who died, like those on this helicopter, were in the prime of their lives: a vibrant part of a family who would never see them again.

When the minister came in on a tide of solemn organ music, I recognized him. He was my neighbour in residence at St Andrews. We had shared a shower room. He too had grown older: more severe and more assured. The service passed, the hymns were sung and *The Lord Is My Shepherd* gave some brief consolation. We moved on to meet and grieve with the families. Jo, who had seen it all before, had grieved too often in this passage through the world of oil. She was strong and sensitive yet conveyed silently that, after the grieving and for the sake of the families, we must go on. As we stood there in the grounds of the church, shaking hands and renewing old acquaintances, I thought of the leaders of the up-stream team and how lucky Shell was to have people like them giving the best of their lives to the enterprise. Somehow as I left I felt this was not the last time we would be grieving here.

We needed to know without delay why the aircraft had failed, and we were assured of regular bulletins. We removed Chinooks from service and would not request replacements of that model, but we had to begin flying again. I was to join the first flight to the platform. We were tense as the safety procedures were read out—we had all heard them before, but this time everyone attended carefully. We began to relax as the flight went on, and in a short time we were on the platform and normal service resumed.

After a thorough investigation it was confirmed that the problem had been in the gearbox, where there had apparently been a modification. Andrew Wylie, chaplain to North Sea Oil and Gas, did sterling work. An exceptionally empathetic person, he had a rare worldliness in his Christianity that disarmed people as he helped them with their problems and their grief. He tended to his flock of about 20,000 workers, which saw him make frequent trips offshore. We were fortunate to have him.

As Tern and Eider struggled to recoup their investment in the face of price collapses, the contractors also started to feel the pinch. They began to bus their employees south rather than using more expensive cars or planes. On a bus there is more time to relax and compare lots, and for the individual to wonder how his might be improved. Before long an unofficial union emerged, and a pressure that had long been absent in the North Sea re-emerged.

When the situation became pressing, I decided to take it head on and arranged a lunch with Ron Smith, the Transport and General Workers' Union leader, and Eric Hammond, leader of the Electricians' Union. I told them bluntly that they were losing membership to this new group. I brought them up to date on our programme of initiatives and the progress we were making, and they said they understood the problem. They had no intention of losing contact with or control of their membership. I promised I would talk to John Brown of BP and brief him on our discussion. John was decisive, easy to deal with and a man of his word. He welcomed the union leaders' cooperative attitude. In the meantime the oil price came back and rose fast, and the forces of supply and demand strengthened, making my platforms attractive profit centres and the work of Smith and Hammond much easier.

It was John, I think, who at one of these conversations told me he had sent two of his men to see Peter Morrison, Minister of State for Energy, to talk about some legislative difficulty or interpretation. They arrived at Morrison's estate in the Western Isles and asked to see him. They had never met him and apparently were a little apprehensive. They had been advised to listen carefully since Peter, a trained lawyer, was at times difficult to follow and in conversation often digressed. They were shown in and served refreshments before Morrison arrived. He was older than they imagined but had a minister's confidence and referred constantly to the 'house', which reassured them. At the mention of legislation, he was off. 'Boundaries must be determined,' he declared. 'Depletion is a real issue, and must be controlled. New stock must be found to replenish the reserves. If you wish legislation to achieve this, Gentlemen, then you shall have it!' And with that, he dismissed them.

The men relayed the conversation to John on their return, saying that although Sir Peter had an odd turn of phrase, he had obviously understood the problem. They did not realise, however, that it had been Peter's father,

Baron Margadale, they had been talking to. 'The house' was the House of Lords where the baron had made only one speech, and that was on salmon and salmon stocks. He had taken the men's visit to mean that somebody had listened to his oration and was planning legislation to deal with his points, and so he told his son when the latter eventually arrived. To his son's enquiry, 'Did you see two young chaps from BP? They should have arrived by now …', he replied,

'No, the only people I have seen were from the fisheries.' It was truly Peter Sellers stuff.

At about this time I was introduced to another baron, John Peyton of Yeoville. As a young man he had been sent to France as part of the British Expeditionary Force, captured in Belgium in May 1940, and spent five years in German POW camps. Having stood down from the House of Commons, he was now, among other things, treasurer of the Zoological Society of London. He invited Jo and me to stay at his home and beautiful garden in Somerset, where he asked me to become a trustee of the zoo. He also promised Jo he would find her charitable funds for the Urban Learning Foundation in East London of which she was a director, and later governor. I did, and he did.

The zoo meetings were mostly about money. The institution was a scientific and research centre trying to become a business. It was important they separated the two functions, but even then the revenues would not cover costs. After a message from the ministry refusing any subvention, John took the bit between his teeth. Irascible is too gentle a word to describe him when his dander was up—and it was up. He addressed the permanent secretary first by acknowledging the ministry's refusal, and then by asking, 'Which animal do I shoot first? I need the answer by Monday because I start shooting on Tuesday. If the answer is not forthcoming, I shall work alphabetically. To avoid any accusations of cruelty, the press will be there to see that it is done humanely.' A cheque arrived the following day, and no shots were fired.

*   *   *   *   *

Since the War, petrol pricing had become a matter of increasing public interest. Freely available supplies of fuel and a huge growth in car ownership had fed public concerns that someone somewhere must be making a fortune at their expense. In September 1960, the government set up an enquiry under the Monopolies and Restrictive Trade Practices

Act to establish whether or not there was a monopoly of wholesale supply. After four-and-a-half years they confirmed there *was* a monopoly, but with some safeguards it was not likely to operate against the public interest. The safeguards concerned the relationships between wholesalers (oil companies) and retailers (service station operators). Time and relative price stability calmed the storm, while continuing growth in car ownership increased public consumption.

The first oil crisis in 1974 changed all that. The rise in crude-oil prices was reflected in the price at the pumps and the government maintained its percentage, taken on that higher price to the benefit of the exchequer. We, the oil companies, now had two major government takes on our price at the pump, but that did not assuage the public anger. A second inquiry was instituted in February 1976. This time the enquiry concentrated on discrimination by oil companies on retailers' terms of trade, the impact of trading stamps and the ownership of stations. Not surprisingly, after two-and-a-half years they found once again that there *was* a complex monopoly, but that it was not acting against the public interest.

The public lost interest from July 1978 to 1979. However, when the Shah of Iran fell from power and his successor, the Ayatollah Khomeini, immediately withdrew two million barrels and more from the market, prices soared, as did prices at pumps and government take, and again public interest brought the House of Commons Select Committee on Trade and Industry into play. In 1987 they instituted an inquiry into the UK's petrol retailing industry, and this was the stage at which I became involved.

The committee made heavy weather of it. They wanted to see the executives who were directly involved, so I fielded my retail manager. After establishing his organizational position they dismissed him: it was rumoured someone actually said that they wanted to see the organ grinder and not the monkey, but I doubt they really went that far. In any event our next submission was the board member in charge of the down-stream sector, a Dutchman and first-class refinery engineer named Jaap. He was also such a dedicated Anglophile that he had taken up fishing and was developing his shooting skills in preparation for the grouse season. It was at that stage he encountered for the first time British bureaucracy in its most insidious form. Jaap returned from his shooting practice one Sunday morning to find two large men at his shoulders as he was opening his front door. 'May we come in?' he understood to be less a request than an order. He had seen too much in Holland during the War not to accede.

As he sat down with one of the men in his living room, the other started poking about his house. Having shown their credentials they proceeded to interrogate him, not on petrol prices as he feared, but on his new sport and the club he had joined. Apparently satisfied, they left.

The next day he related this to me and a couple of fellow executives who were involved in government relations and security. It took one of them just 24 hours to discover that Jaap was the only non-Irish member of the shooting club. He was the 'Jackal' suspect number one. This could have made the Select Committee an even more daunting experience, but Jaap took it in his stride and performed brilliantly.

The committee opined that the matter required further investigation. They went on to recommend that the Monopolies and Mergers Commission should investigate the petroleum retailing industry in the UK. In late 1989 the case was referred.

Sydney Lipworth, the chairman, would head the inquiry. When the day came we sat there in rows with the press in the majority and lawyers in abundance. Sydney opened proceedings by asking me to attend a private session in his chambers. In a hushed auditorium, I duly rose and followed him out of the room. I had no idea what was on his mind, nor any prior indication that we had done anything that might have disturbed him. As we sat down he said in an unusually nervous way that he had a huge apology to make. The confidential list we had supplied detailing our prices to all our commercial and industrial customers had, he told me, by some terrible mistake been circulated to our competitors. This was astonishing, and for a moment I was stunned into silence, but I was not going to climb onto my high horse just then. That would have been an unfortunate start to what experience told me would be a lengthy process. I said, 'Sydney, in this volatile world what is fact today will be history tomorrow—so let's get on with it. This is a competitive market as you will find out, and all prices are fluid.'

The enquiry looked at prices, profits and the agreements that underpinned the devolved retail structure. They found that the level of prices was not unreasonable considering the movement of crude-oil prices and consequent government taxation. They examined and dismissed the accusation that prices were raised too quickly, and the concern that since price increases moved together there could be collusion. Finally, they examined down-stream profits and found them to be below the level of industrial profit in the UK.

Overall this was a fair report. The examination under Lipworth's direction was thorough and the process painstaking, and it was reassuring for the public that such an examination had taken place. For us it was neither a victory nor a justification, it was a reality that we had to continue to communicate with confidence. For our staff and contracted retailers it was a relief that what they had been telling their customers was now confirmed and underwritten by the highest authorities in the land. There is, however, no doubt in my mind that it will come up again, such is the public aversion to paying large sums to keep moving forward.

*   *   *   *   *

Our safety performance continued to improve but the elements—the winds and storms of the North Sea—were the ultimate challenge. A gusting wind blew a helicopter off a platform landing pad as it was about to discharge its human cargo. It fell into the sea and sank. The training that was now compulsory had taught the occupants to stay in their seats until the craft settled on the seabed, using the air that was locked inside with them. Once the craft had settled, the drill was to push the windows out and then rise and swim to the surface. This time all the occupants except the two who had been killed on impact made it to the surface and were rescued. It was a dreadfully sad event, but would have been far worse without the training.

A platform has a drilling capacity and a production system that controls the separation of oil from gas, and then discharges the outputs into discrete pipeline systems for shipment to shore. Any work undertaken on the production systems must be authorized with a permit. This is an essential control and stops any work taking place on the system if there is already activity ongoing. Work can begin on a new permit only once the old one is completed, so vulnerability is limited to one job at a time.

The issuer of the permit has a full understanding of the production system: the layout, the connections and the valves controlling pressures and flows. Major platforms have the capacity to accept flows from adjoining platforms for processing and transmission to shore. While this potential can offer substantial financial advantage, it also complicates the purpose of the original platform, making the issue of the permit more complex.

As the day-to-day work went on with all its risks and challenges, I contemplated how little the end users know about what is involved in bringing light and warmth into a home. But that changed dramatically in 1988 when film of a sinking platform and a raging inferno was flashed around the world, and a horrific explosion caused the deaths of 167 people.

Piper Alpha was a North Sea production platform 120 miles north-east of Aberdeen. It operated initially from its own production, but became a 'mother' platform for oil and gas production from the neighbouring Tartan and Claymore platforms. These flows did not switch off when Piper Alpha's situation was deteriorating and fires had begun in the pumping area. The fires' intensity was so great that access to the fire pumps was impossible, and they did not activate. When the burning gas-oil flowed under the incoming gas line, it split like a peeled banana, injecting millions of cubic feet of gas into a thunderball.

*   *   *   *   *

Lord Cullen, a contemporary of mine at St Andrews, A dux (top pupil) of Dundee High School, an eminent scholar then lawyer and law lord was appointed to hold the enquiry. He opened it with a roll call of each individual who perished in the tragedy. This was as salutary as it was solemn.

The inquiry tracked the tragic course of events and examined the safety arrangements, practices, decision-making and planning that had led to a programme of activities vulnerable in ambition and seriously exposed in execution. The North Sea's subsurface elements have immense power. For centuries fishermen from Aberdeen had battled the waves on the surface and the gales whipping across them. Now, from castles built on the seabed, engineers tackled the forces beneath, confident in their technology until they overreached themselves and were swamped, like many of the fishermen before them, by a catch too powerful to land.

Cullen's legacy was the safety case, the thought-through analysis of the risks involved in any industrial activity, be it a railway, an electricity generation plant or an oil platform. This analysis laid out the processes for managing the risks and the steps that should be taken to mitigate the consequences of any unforeseen incident. Within a few years, each North Sea platform had its safety case approved by authorities. Years later I was agreeing safety cases for the new railway bodies under privatisation. This, as I told my staff at that time, was security for the future: analyse your vulnerabilities, mitigate them, and if you are still not satisfied, redesign them until you are. It may cost the taxpayers, but it is in their interests.

The tragedy of Piper Alpha continued to dominate the press for many weeks. In London, where events tumble on top of each other day after day, a headline can be in the first edition of the *Evening Standard* and lost in the small print by the last edition. Not so in Scotland, and not so Piper Alpha.

*   *   *   *   *

Not long after the accident I received a notice from Aberdeen University advising me that they were awarding me an honorary degree. At the dinner on the eve of the ceremony, the principal took me to one side to seek my advice. Apparently newspaper proprietor Robert Maxwell, who was also to receive an honorary degree the following day, had offered the university a large donation for the creation of a Department of Accident Prevention. He was also planning to open an appeal in his paper. Maxwell had a practice of opening appeals in good causes. My first reaction was that there was not a body of knowledge entitled 'Accident Prevention', nor was it appropriate for a university in such emotional circumstances to launch such a department. The Cullen Inquiry had not yet begun; indeed memorial services were still taking place. I warned the principal that such involvement would appear opportunistic and the university would be unwise to be associated with it. I undertook to convey the message to Maxwell that evening before he arrived. 'But what about the money?' the principal said.

'I suggest you forget it. You have just avoided creating a non-department with costs and no income to offset them.'

'What shall I say to him?'

'Nothing,' I said. 'He won't mention it.' And so it was.

The next day Maxwell arrived in his new private plane, shook hands all round, received his degree and sat down to lunch, in the middle of which he announced that he had to fly immediately to Paris. He disappeared, leaving his wife with no evident means of transport. In the end we gave this most engaging and pleasant woman a lift home, leaving a bemused principal behind. The rest of Maxwell's life is unclear history.

*   *   *   *   *

Cabinet minister Cecil Parkinson was a golfer and suggested a game with me and Willie Ricketts, a civil servant in the Cabinet Office who was working on electricity privatisation. It was a fine day and the golf was good. Willie's description of the plan for, and progress of, privatisation was fascinating. At the end of the round Cecil asked me to come and see him next week. I agreed; we were due to meet in any case so I could brief him on the final stages of the Monopolies Commission report.

When I arrived and before I had had a chance to brief him, he asked if I would take on the running of the railways. He described the aftermath of the 1988 Clapham Junction disaster, and his concerns about morale and the state of the network, and said he believed my experience could bring much to the role. Privatisation was not being considered then, he said: it would be too difficult, too complex and would have no public appeal. I asked if Mrs Thatcher backed his approach, and he said that she did.

I reminded him that I had come to brief him, not to solicit a job offer, but said I would consider it and come back to him within three days. My family were not uniformly enthusiastic. Douglas, our eldest, said it was a poisoned chalice. Jo said, 'Whatever you want to do, do it; but if you don't, don't regret it.' Paul, our middle son, was more bullish: 'You haven't had a job you couldn't do so far, so why don't you take this on and see how good you are?' This I took as a 'yes'. Michael, our youngest, liked the idea. I replied to Cecil that I would talk to Shell.

Peter Holmes (now Shell's chairman) was not delighted with the news, but agreed that any further development for me would have to be back in the Towers of Contemplation and Strategy, since at my level the rough and tumble was no longer available. The prospect was not attractive compared with the challenge of a 135,000-people railway trying to recover from a tragedy and a destructive strike. These were situations I well understood.

Being a man of the harsh and bitter desert (a mountaineer and explorer, Peter had a military bearing and a detached, analytical mind), he then took his pound of flesh. I would have to cover the job at Shell until mid-year, he said, in spite of the fact that a successor was already keen to step into the role. This would mean three months of doing both jobs at once.

*   *   *   *   *

Leaving the company was a difficult, emotional experience. At the last farewell party at the Victoria & Albert Museum, a good friend pressed a parcel into my hand. 'Thank you, you've done your job,' he said. When I got home I opened the parcel to find a picture of a man leaving the sea behind him, his footprints, deep in the sand, being washed out by the incoming tide. Such is the course of human affairs, the picture was saying—in Nigerian terms, 'No condition is permanent.'

As I sat among the pine trees that weekend listening to the birds, I realised that the disappearing footsteps in the picture reflected no sadness for me. I had experienced every minute of my life at Shell to the full: its joys, its sadnesses, its progress. As I cast my mind back to where we had been and what we had been through, I could only thank those who had given me and Jo the opportunity.

Borneo had given us the first taste of impending trouble, exaggerated at first, perhaps, but then exploding into insurrection. Nigeria was a dramatic 14 years of history, from a pomp-and-circumstance handover of colonial power to a downward slope of financial and political corruption, brutal slaughter, pogroms and bloodshed, and finally a civil war that pitted tribe against tribe.

After the forces of resurrection eventually took hold, we exchanged a quieter Nigeria for a Thailand surrounded by conflict and revolution. Each of their neighbours' situations passed through their violent tribulations into states of exhausted equilibrium while Thailand itself wove its way between the potentially destructive obstacles they had put in its way.

The missions to Iran that followed brought to mind at least one historical echo: Henry VIII taking leadership of the Church, thus facilitating his massive land-grab from the monasteries. Strategic distribution of those lands subsequently gained long-lasting public support for his regime. By contrast the Shah of Iran, in conflict with the Mosque, did not have this deep-rooted support, and when the Bazaar (the merchant class and market workers, the traditionalists) turned against him as well, he was gone; Khomeini and the Mosque returned. Iran did not find immediate equilibrium with its new theocratic system of government, and indeed is still to find it.

Australia was another totally different experience. Here was a country where, had I been in my mid-20s, I might have settled. It is a proud democracy with a people who have developed the land and gained wealth from its natural resources. Their achievement underpins their confidence, but over decades, industrial relations tussles have scarred management and unions alike. This is not a battle that can be or has to be won, and happily Australians are beginning to see this.

Looking at Australia as a homeland, which it is for two of my sons and their growing families, I also see a haunting sadness. This has to do with all those other sons lost in the fields at Gallipoli, and the battles that followed in the jungles of Asia in the next war.

My success in these contrasting situations depended largely on the leadership skills I could bring to bear. From the start, my first preoccupation was with the people, their selection, their development and their training. Because of my early immersion in an international business and perhaps also my anti-sectarian upbringing, I was at ease with race, gender and background, so immense, unrestricted pools of talent and support were open to me.

My second preoccupation had been with performance, and here safety and quality ranked highest. I had learned that the hard way, aged nine. Close behind that came the in-depth understanding of proceeds: your costs and your cash flow. Pricing sausages, discounting turkeys and collecting bills from drunken workmen gave me an early understanding of the grass roots, the basis of any successful commercial venture. Building on that and being open to opportunities were lessons amply illustrated by Shell's founders, who had succeeded in creating one of the greatest companies in the world.

My third preoccupation had been to build challenge into my management teams to encourage debate, define options and help find the best way forward. You do not need to like every member of your team, but you do need their individual contributions.

It was a privilege to have enjoyed the kaleidoscope of experience in these contrasting and exciting circumstances. To have shared it all with a person of high intellect, independence of mind and enormous empathy for the people—particularly the women—involved with us in these situations, was an inexpressible bonus.

9

British Rail: On Track and Off

British Rail's outgoing chairman, Sir Robert Reid, and I already knew one another, so on my first day the preliminaries were brief. He made sure I was informed about the industrial relations situation, the management I was inheriting, the network and the level of under-investment, and strongly advised me to go out and look at the infrastructure for myself. Sir Robert had run it with skill and knowledge acquired over many years, so—unlike me—in searching for solutions he knew where to look and what to ask. His prescient parting words were: 'Good luck!'

The first thing I needed was an executive team that focused the responsibility for, and control of, all the activities. After a discussion with chief executive John Welsby, we put the team together. For finance we had James Jerram, a constructive contributor across the range of relevant issues; for operations, David Rayner, widely experienced and with a sound feel for industrial relations. For strategic planning we had John Edmonds, and he was the only one I knew anything about—Peter Holmes had shared a room with him at Cambridge. As always, Peter was acerbic: 'He's introverted, very clever and besotted with railways—but watch him: he is clear in his objectives, and determined.' The personnel function needed attention, and there was a gap in engineering and technical matters, but I would fill that soon. The board secretary, Peter Trewin, was a gem: knowledgeable in the extreme and a safe pair of hands.

My two first challenges were the Clapham Junction disaster and subsequent report by Anthony Hidden QC, and the disastrous strike, which had ruined what remained of British Rail's reputation. Understanding the strike was key to assessing the effectiveness of the personnel directorate and might also help to explain the underlying cause of the Clapham disaster.

The strike had been the outcome of a failed negotiation over wages. The car industry had attempted to solve its union relationships with a sophisticated restructuring of its wage package that required the union first to understand and accept the changes, and then sell them to the rank and file. This approach had had a positive press and had an attractive feel about it: modernization, delegated responsibilities, technical education and so on. This was not the railways, however, and our personnel team did not see it working for us—nor, in fairness, did the board. At this juncture Paul Watkinson, a realist who was in touch with key union members, stepped in to take over as personnel manager.

Engineer Peter Watson completed the team. He had worked with GKN, a multinational automotive and aerospace components company, first in research and development and then as CEO of two of its operating divisions. I was very satisfied with this group. They were pro-active and motivated to push for continuous improvement. I could trust almost every one of them, and where trust might be wanting there was flexibility and a creativity that we might need as we set out to recast our future.

Marion Chapman looked after the chairman's office. Severe, slight and quiet, she was also knowledgeable, decisive and exceptionally determined. She had been brought up an orphan in the Midlands, and as soon as she could she had set off to London to develop her rudimentary typing skills. She ended up in the railways as an assistant in a manager's office. She had been in the job for only a week when the train she was travelling to work on was involved in a high-speed collision. She extricated herself from the wreckage and, resolved not to be late in to the office, continued her journey by bus and on foot, managing to arrive on time. Within minutes Marion was called in to an emergency meeting to take notes. Her boss told his colleagues what had been reported to him of the accident. When he had finished she piped up, 'It wasn't like that,' and proceeded to describe what had actually happened. Amazed, they asked her how she knew all this, to which she replied: 'I was there.'

'So how come you're here?' they asked.

'Because I had to be on time.'

In railway lore this was legendary stuff. With that same tenacity she had reached her now hugely responsible role. Soon after I joined, Marion had been confronted by two press photographers who had cheated their way into my office to take photographs of its redecoration. She calmly called the Transport Police and had the men held in custody. When I arrived later, I rang the newspaper's editor and arranged to meet him for lunch. I explained that while the escapade had been stupid, the consequences could have been far more serious. We were at that time shipping armaments across the country in special trains to be despatched to the Falklands, so any and all material in the chairman's office could be considered confidential, if not a matter of national security. He was truly apologetic. We went on to talk about the railways, our plans and challenges, and from that day onwards we got a good press from him. It was also reassuring to know that my office worked efficiently and unflappably, and would not be messed about with in the days to come.

The Clapham Junction crash had happened almost a year and a half before I joined, but it was still current in people's minds. It was an appalling incident: 35 people were killed, many more injured and multiple trains involved. *The Hidden Report* about it was published in September 1989. The cause of the accident was incorrect wiring. A redundant wire had not been removed; it remained connected at one end and bare at the other. The loose end came into contact with a relay and this turned a red signal green, giving the train the OK to go forward into disaster.

Hidden's analysis was incisive and fair and did not seek to apportion blame, which was a relief to both management and board, but eight months after its publication and 18 months after the disaster, the accident still hung like a pall over the entire staff. It was essential to act, and to act swiftly.

First we had to deal with the High Court case. I immediately quashed the suggestion that this was management business. This was board business. When a corporation is charged, it is the board and the chairman who should answer. I outlined how I would make my statement to the court, what would be in it, and how deep apologies should be recorded. After a discussion on admission of guilt, it was agreed.

Next we tackled the signalling staff's working schedules. Weekend working was then the norm and would be compensated for by days off during the week. We would reorganize this so that Saturday and Sunday would effectively be the same as Monday to Friday. Overtime and double overtime would not apply, but the basic wage would be increased significantly to compensate.

Any death on the railway would require a full report to the executive committee, written up and presented by the senior manager. We would re-examine our training programmes to ensure they focused on our vulnerabilities.

The court case came and went. I made our statement and we paid the price. Although this closed the matter in one sense, the safety issue as far as we were concerned was only just beginning.

John Welsby and I tackled Jimmy Knapp on the signallers' issue. Jimmy was head of the National Union of Rail, Maritime and Transport Workers (RMT), the major union involved in day-to-day operations. A straightforward Scotsman with one of the best-known voices and accents in the country, Jimmy was almost invariably a man of his word. As our acquaintance developed into friendship, I learned that you could discuss matters with him in confidence and they would go no further. This proved vital in the years ahead.

The signallers made up Jimmy's power base, so he was highly sensitive to the wage issue affecting them. Our proposed increase in basic pay in return for a radical rearrangement of the working week would significantly enhance the signallers' future pension entitlements, but Jimmy was reluctant to agree, and ultimately negative. I put a time limit on it, setting out in a letter the pension implications, which included improved widows' pension entitlements. The union was not too happy—this was a protective consensus organization—but the letters went out directly to the individuals' home addresses. This approach had worked like a charm for the State Premier in Queensland, and so it was with us. The wives saw the point and within a couple of weeks we had over 80 per cent acceptances. Jimmy accused us of 'dirty pool', but in truth I believe he was as relieved as we were. *The Hidden Report* bore as much on him and his union as it did on us.

The cause of the accident revealed a weakness in supervision and a basic lack of preparation for a critical piece of work. It was just as we had found with the work on platforms and in refineries: you cannot set up safe work programmes until you have assessed your vulnerability. The discipline of this process helps create a safe work place. None of this had been implemented at Clapham.

I had undertaken to visit every site where an employee or contractor was killed, and to meet the local manager, supervisor and workforce. Even as the safety reporting to the executive committee got under way, an employee working on the down-line out of London was hit and killed by a fast train. An excellent worker, he had apparently already examined the problem on the line and was working out what to do about it. There had been a strong wind blowing sound away from him so, concentrating on the job in hand, he had not heard the train approaching until it was too late. His determination to get on with the job had led to his death. Here again there had been no catastrophic vulnerability assessment, so no safety warning system set up.

The training course initiative got under way, and how to establish a safe working site was top of the list. Initially the interaction was poor—the participants were there to be told what to do, not to contribute. The existing book of regulations had two apparent objectives: to be comprehensive and to place the responsibility for safety on the worker. It was not written in plain English, so understanding it fully required a high reading age. It offered no support from the managerial structure: if anything went wrong, the buck stopped with the worker. As a practical guide to safe practice on a railway, it was useless.

As the week passed the training groups began to participate, rather than just listen, in seminars. Sadly they had real incidents to examine as the death toll grew and the improvement engineering works got under way. Men were out at night in the pitch dark, sometimes in strong winds and bitter cold. The risks were high. Young technicians looking for their colleagues or locating the work site were walking up and down the line as opposed to alongside it at a safe distance. In these months we on the executive committee were learning how our railway was actually running and how much we had to do. It was all about standards and self-discipline.

Then it happened. Near Liverpool, a three-man team was out to fix a failed signal on a dark, wet night. As they walked past the signalling centre they did not stop to check—they had their orders and they were going to carry them out. They ignored the path next to the railway: it was uneven and slippery and you could turn your ankle. A torch would have made a big difference, but they never used one. Had they done so that night, things might have turned out differently. Meanwhile, a massive piece of machinery was at work churning the stones that lay between the rails; they had to be turned regularly to prevent them from sinking into the soil. The noise was overpowering, dominating the night air to the exclusion of any other sound, so the men did not hear the freight locomotive as it bore down on them, then left them, invisible to the driver, dying on the track. By this time the ballast train had turned and was at work on their line; within minutes it was upon and over them. They died twice.

I heard the news at 3.30 a.m. and was on my way before daylight. By mid-morning, I was taking the minute's silence on site. The team was devastated; each one had been a personal friend of the three men. We all sat down in the changing room. I talked of death in the workplace first as something that had happened and could not be undone. I spoke of the oil business where the accidents and disasters all have a cause, and how the pursuit of that cause has to be relentless. The lessons learned have to inform and educate for the future. So it was with the railways. We had to learn and we had to change. The mood was beyond sombre.

On my way home, I struggled with my emotions and asked myself what more I could do. As I had sat with the team, the hopelessness had seemed overwhelming. The barbarity of a double death was horrific. Clapham had been a symptom; the deaths on the line were the disease. If we could not secure ourselves, how could we secure our passengers? We had to drive the safety message home; nobody could do that for us.

We had returned to the training and to spreading the message when something happened that was a huge boost to morale. The signalling team made a film called *Dead Serious about Safety*. It was narrated by one of the widows. When the board saw it, there was a stunned silence. Because it was the grass roots talking, the message was clear and relevant. Assessment of vulnerability and the creation of a safe workplace was finally becoming a reality, just as after Piper Alpha, platform operations' safety assessments became the norm. Incidentally, the film went on to win the Audience Favourite award at the first European Video Film Festival for Occupational Health & Safety.

Businesses—or more accurately, organizations—have a soul. Staff are more than boxes with titles and names; they are people with hours, days and years invested in them. Confronted by extraordinary circumstances, they react in unexpected ways: the Dutchman who had protected his assets as if they were his personal legacy but when it counted said, 'We can replace boats, but we cannot replace lives. Let them go'; the secretary who stepped over the wreckage and brushed aside the trauma to get to work on time; the widow who came out of mourning to honour her cruelly killed husband. In the railways, as in Shell, the soul was there and there were people to protect.

The accident-prevention performance was improving; preparation for the Channel Tunnel route had knocked down 89 bridges and there had been a single death at the end of that programme. One too many, of course; but had the new procedures not been in place, there would almost certainly have been more. The numbers of deaths on the line had dropped markedly. The message was getting through to a receptive workforce. Responsibility was not for allocating elsewhere but for embracing.

*   *   *   *   *

After safety, the next issue troubling Parkinson and Mrs Thatcher had been the state of our industrial relations. Battles with the coal industry, while achieving a victory for her, had drawn into the fray not only the secretary of state and his minister, but also the prime minister herself.

As a first step I took a careful look at what happened in 1989's National Rail Strike. Reading the sequence of events, emergency meeting after emergency meeting, it was clear that confusion had reigned. The structure around the negotiation had lacked clarity. Involvement widened from the railway executive to the board to the Department of Transport and then to the Treasury, who were vocal on the financials, making a deal almost unreachable. It had reached the Cabinet Office, where constituency frustration was vented and so the possibility of a tidy deal receded. Exhaustion and lack of money finally won the day.

My first industrial relations negotiation was rapidly approaching. John Welsby and I discussed our proposed approach and agreed there was to be no proposal for complex and sophisticated organizational change as was attempted in 1989. This negotiation would revolve around the retail price index. Paul Watkinson was to lead the negotiation. There would be a small directing group including Welsby and me, and I would keep the board and secretary of state informed as negotiations progressed.

In 1989 the railways' operational management had been neither involved nor informed as the debacle made its way to its sad conclusion. As a result there was no opportunity for positive communication and consultation with the workforce. I was concerned to find that the supervisor segment of our workforce was neither union nor management. To my mind this was a serious fault that needed to be addressed.

We prepared for the negotiations with a comprehensive briefing programme that encouraged support for a positive outcome. This was especially important for those in the supervisor role, the interface between union and management. At first this approach seemed strange to middle management, but they quickly understood and welcomed it. The preparation paid off and the deal was done, somewhat to the surprise of the politicians, but I warned them there was more to come. After the negotiation was over I sat down to take stock.

We had an aging workforce. By the turn of the century, 70 per cent of our drivers would have to be replaced. Most of those leaving us would have had a long apprenticeship and 40 or more years' service. This was a solid platform of competence and each driver I met bore that out.

The railway operators did not have the same depth of experience but nevertheless they were highly capable, as were the engineers. To this strength you could add experienced managers and supervisors. So you might fairly conclude that the erratic performance reflected the state of the assets, not the competence of the people. Employing more people to improve performance would result in low productivity and disappointment to both the organization and the customer. We urgently needed a process by which we could match resources to need. This in turn would lead to effective operation, because in defining the need you would also define the most efficient training and development programmes.

My predecessor's concept was to devolve the running of the railways to cost-effective organizational units which addressed customer requirements such as capacity, frequency and the quality of the provisions supplied. The attempt to introduce this idea in the 1989 negotiations failed, but I felt that this aspect of Sir Robert's plan should be implemented.

In establishing these new units I was keen to change from a command culture to a participative, contributory way of working that allowed individuals to grow with the job and support their organization. I was also concerned that our relationship with the union was blighted by confrontation. There was no way you could run an industry that had well over 100,000 people involved without improving this situation. This was enough of a challenge, but its urgency was increased by the spectre of privatisation. Parkinson had told me before I joined that privatisation was not on the agenda and Mrs Thatcher, suffering the aftermath of the Poll Tax imbroglio, was of the same mind, but as both these figures disappeared I was not so sure. If we *were* to privatise, we would need to be in much better shape.

The Associated Society of Locomotive Engineers and Firemen (ASLEF) held a weekend conference hosted by General Secretary Derrick Fullick. Jo and I went to join them. The conference was excellent, the informal part even better, and we both learned a lot about the drivers' lives and families. They were an exceptional band of people. I presented my objectives, stressing safety as a priority. I developed the theme that the future of the railways in whatever form depended on investment in assets and in people. One point in particular seemed to make an impression. I had visited the mayor's parlour in Peterborough's town hall where photographs of past mayors were displayed, and noticed that almost all of them had been train drivers. And Peterborough was not an exception: the same correlation could be found in Birmingham and in London. Why, I asked them and myself, had this development of management talent happened outside the railways, and not inside?

*   *   *   *   *

Operational executives Ivor Warburton, Chris Green and Brian Burdsall had become involved in a philosophical analysis of the forces that change human behaviour. They were all experienced managers who could relate to the need for change, and more importantly they believed it could happen. I personally was enthused and excited by what they were working on: this was something that was going to change the railways.

The origins of what became a movement were rooted in the 1940s in the US, when W. Edwards Deming had made a substantial contribution to improving productivity. His efforts on war production led General Douglas Macarthur to invite him to Japan in 1950 to work with the leaders of Japanese industry. Japan's energetic acceptance of the importance of quality drove their huge success in the post-war world, and Deming was honoured for his part in the renaissance. He summarized his philosophy as 'reducing variations'. Over 90 per cent of production problems, he believed, arose from system irregularities and inefficiencies rather than being the fault of those operating the processes. He also believed that those participating in the process must contribute to its improvement. This called for a step-change in management style, which is what had excited our managers. We could no longer simply allocate tasks; it had to be a contributory, collaborative process with everyone working together.

Deming's thesis involved removing the blame culture so that mistakes could be exposed without fear of retribution. Mistakes were there to be learned from. Another engineer/management consultant who contributed to Japan's resurgence, Joseph Juran, promulgated the 'fit for purpose' thesis, out of which was born the concept of quality planning and control. This was the intellectual basis for our next major step forward: organizing for quality.

Peter, now technical director, supported the works of Juran and Deming, but he was also an adherent of *kaizen*, the Sino-Japanese word for improvement. In business, it refers to activities that continuously improve all functions and involve all employees, from CEO to line-worker. The theory advocates incremental change: do not patch and mend—build new and fit for purpose. Watson's experience and enthusiasm helped him push the change through at board level, and executives welcomed the way their initiatives were being moulded into a quality revolution. This would change the railways, and despite the unforeseen diversions that lay ahead, we had clear public evidence of its success at the first Quality Exhibition at the Birmingham Exhibition Centre. The exhibition showed that teams were tackling chronic problems in a cross-functional way, and solving them.

Improvements generated by these initiatives stoked the fires of change. One outstanding example involved the so-called Misery Line, which ran from London to Southend. Its statistics justified the name but, working together, teams saw that the root cause of the delays lay in the positioning of trains for the evening peak. They relocated them to the depot closest to London, instead of in Southend where they had been. The change—so obvious in retrospect—made an immediate and spectacular improvement.

Putting past and present right could not be enough and so we developed *Future Rail*, a document that laid out plans for modernizing the railways. Channel Tunnel preparations were the top priority, followed by further improvements to the Misery Line which would cost £350 million, and replacement of the antiquated Fenchurch Street signal box, which in its current state could have caused a disaster of Clapham-like proportions. The modernization of the Kent line to the tunnel, and the building of the station in Waterloo to accommodate the new line were also key features.

The schedule was demanding. We organized a progress and decision-making meeting at 8 a.m. every second Friday. There were no excuses; the finance management was there each fortnight to sign off and pay for work completed, which motivated the contractors and suppliers.

The programme was on target—until the first trial run of the TGV in the UK. The new train, on entering Dollands Moor Yard, turned the red lights in the neighbouring station to green. It was a huge setback. Powerful as it was, the train did not have enough space on board to accommodate equipment to suppress this risk, so we had to redesign and re-equip our line to overcome the problem. It was a tribute to our engineers and physicists that we managed it, and on schedule.

Privatisation was now a high priority to Tory Transport Secretary Malcolm Rifkind. Increasingly alarmed at his apparent lack of understanding of the railways and of the difference our improvement programmes were already making, I wrote to him at length. I drew his attention to the fact that, though the route to privatisation was being formulated along traditional lines, the idea of a central authority dispensing franchises simply did not pass any normal test of viability or operability. The government was concentrating solely on the transfer of ownership. How much better it would be, I argued, if they saw it as a progressive introduction of private-sector capital. This would avoid the unbelievably costly and time-consuming bureaucracy, and would also allow our improvements to continue undisturbed.

By the time the discussions began, the Channel Tunnel and huge refurbishment programmes were already under way. We had our capital expenditure at over half a billion pounds a year. This required focus and undiluted management attention, but for a politician, I regret to say, that was another irrelevant consideration. I asked for a clarification of their objectives. It was easy to define what they were not: not completion, not quality, and not a massive return on invested capital, because there would have to be a huge discount on the national assets that were to be sold. This seemed neither clever politics nor sound financial arithmetic.

As the heat intensified and with an election imminent, I made a decision much influenced by Jimmy Knapp's well-thought-out position: we live in a democracy, the railways belong to the nation, and it is for the elected government to run or dispense with the assets as they see fit. We would therefore not frustrate their will, provided it was safe. It would be unwise, unfair and confusing to my employees for me to initiate a campaign to frustrate a legitimate authority. I would make this clear at every stage.

The election came and went and we arranged a meeting with the new transport secretary, John McGregor, so we could brief him fully. John had been at St Andrews University, and had possibly even been a student of Jo's while I was in Borneo.

We set out first of all to place the railways in the context of the country's economic life: 762 million journeys a year; 16,000 trains a day; 500,000 (and rising) commuters a day into London and a quarter of a million into other regional conurbations; 75 million tons of coal moved each year and over 500,000 containers on a network of 2,500 stations and 8,000 bridges and viaducts. We compared British Rail's operating costs of £8 per train kilometre with those of France at £13, Germany (this was before Unification) at £15 and Italy at £22; with financial support of 0.12 per cent of GDP as opposed to at 0.61 per cent in France, 0.56 per cent in Germany and 0.92 per cent in Italy.

Having demonstrated that we were an integral part of the nation's economy and society, and argued that we should continue to be so, I described our proposition of bringing private sector investment into the railways. First we would move from our monolithic engineering structure to a contracting model. Standards and processes would be protected by a tightly managed unit that would contract out work to engineering constructors. This would ensure competitive costing and pricing, and help develop British firms that could win and build international businesses.

It had been a long morning and the presentations were good. At lunch we met the Rasputins who surround our Czars. The spokesman, pontificating on command and control (neither of which he had experience of), produced a series of analogies to illustrate his case for privatisation. First it was with the airlines: easily dealt with as they have open skies and no single rails. Then the electricity industry—which, unlike the railways, has captive customers who must switch on if they want to see in the dark or stay warm. Our customers by contrast could catch a bus or even use the telephone. Each analogy was shot down, but this did not deter them. There was money in it for them!

After lunch we talked about investment: how the railways had been starved through the '80s and how the Channel Tunnel investment should be seen as an addition, not a substitute. I pointed out I had now accepted budgets, but not agreed them. The department knew that and so did the treasury. The simple fact was that the state of the railway's assets was unacceptable. In his planning McGregor should remember this and build enough into the allocation to put it right. This should be his number-one priority ... I bit my tongue here and managed not to say 'before embarking on any political adventures'.

Turning to his manifesto position and the concept of franchising, we showed how we would implement this, making it clear that we were not opposed to change. The franchise would be developed out of four new organizations set up by the relevant profit centres. This would keep the production management in place and ensure smooth running. We defined Network SouthEast and Scotrail as first steps. The franchise would be a retail-serving one, covering stations and car parking.

We turned then to the proposed issue of access liberalisation—allowing other train operators to use the infrastructure. We started with the reality: we had a congested system with carefully designed allocation paths. To disrupt this for minute movements and to disturb the architecture of decision-making (which as it stood was transparently fair) made no sense at all. Subsequently the department commissioned consultants to analyse the feasibility of access pricing. The information requirements for this task were enormous and the information collected and processed had nothing to do with running the business. The study ran into the ground when they discovered they could not attribute more than 18 per cent of the direct operating costs to individual train movements. Money ill spent! It also became clear that open access would require intrusive regulation, which would enormously increase the management task, and therefore cost. Already operating in highly competitive markets with negative profitability, the railways could not afford this extra burden.

Lastly we turned to freight. In 1976 we had 14 per cent of the freight transport market. By 1990 this had fallen to 7 per cent. The predominant competitor and supplier was road transport, which operated without paying the true costs of its infrastructure and could offer infinitely more flexibility with regard to schedule and route, including door-to-door delivery. The railways held the advantage in the markets of coal and stone. Where the railways ran from source—the coalmine or the quarry—to the point of use, the power station or construction site, it was viable: these were stable contract markets. Why disturb them for political presentation, particularly as their economics were relatively unattractive?

This ended a long day. As we sat discussing the meeting afterwards, gloom descended. The questions we had been asked betrayed a lack of concern for the practical, and a bias for the bureaucrats' solution. The operation's fast-improving performance seemed to count for nothing in either their calculations or their decision-making. Whether or not we played a blinder, as reported, did not matter. The caravan was on the move, to a destination chosen by the people's representatives.

On my own later, I weighed up my options. I could rubbish the announcements and withdraw, or stick with it and continue to try to mould this into a workable shape. The decision did not take long: my loyalty was with the people of the railways, and I would stay with them. I would not fight to frustrate and deny the political will—although if I did, I would not be alone. But such a course could be chaotic and bloody and would get nowhere; it could also wreck many promising careers in its wake. From where I was standing I still had the chance to argue out the more impractical—and frankly idiotic—parts of their proposals behind closed doors.

❋   ❋   ❋   ❋   ❋

By 1992 our 'matching resources to need' programme was under way, and was already producing significant levels of surplus which we advised the unions would be handled in the first instance by a programme of voluntary redundancy. By the end of the year, we had 10,000 employees opting for the programme, and this number grew as the initiative continued.

Our case to the public was based on the state of our finances, which had been badly hit by the recession. The surplus of £18.2 million in 1988–89 on our day-to-day activities had been transformed into a loss of £93.1 million in just two years.

As we struggled to achieve viability by reconciling costs to revenues, Jimmy Knapp and his executive could see their power base shrink and were determined to make a stand on redundancy. We were back into strike mode again, but this time we were better organized. Our managers and supervisors understood there are no jobs for life. This resonated with the media—one of Jimmy's normally supportive journalists wrote that not even in Japan were there now jobs for life. Jimmy's strength lay in his tactical thinking and his understanding of the possible. The weakness of his position lay in the structure and attitude of his executive committee, who had no reason to reach agreements quickly. This would be a huge handicap in a fast-changing world.

Discussions were heated but finally Jimmy accepted a deal that kept our initiative intact, and we arranged to meet with his executive to endorse it. In front of the 'hard men' of his executive the following day, however, Jimmy presented his disagreement with all that had been accepted the previous day, and proceeded to table a whole new set of issues and demands. For me, the turning point was his demand for a guarantee of 'no compulsory redundancy'. This, I told them, was a political agenda, not an industrial relations agenda. We were in a new world of change and I advised them we were withdrawing the 'check-off'. This was the system by which the union collected £500,000 a month from their members' pay packets. This created uproar, and I was accused of being abusive and a bully.

The meeting lasted only eight minutes. As we emerged, the waiting media asked Jimmy if the meeting had been short and sweet. 'Only short,' he replied curtly. The *Evening Standard* was not alone in wondering why I seemed to have acted out of my normal 'cool, calm, collected' character. The answer was simple: we were at the end of an era and they had to understand it. The stakes were too high, especially if privatisation was on the cards. If we were to maintain any form of discipline and order we had to make it clear that the industry had firm leadership. Moreover I was not prepared to reach what was virtually agreement with the general secretary one day only to have it undone by the executive the next. In simple terms, they were not going to succeed where Jimmy had failed. They needed a severe dose of reality and it was their actions that had lost the union its check-off. They knew this, as did everybody else, and it was salutary.

ASLEF realised the seriousness of our position and took themselves out of the argument. With them out, we set out to reach a settlement with Jimmy. It became clear they could not raise the votes to sanction another strike, so they went for a ballot. Once a ballot is called there cannot be a strike until the result is announced, so we set to work to win the ballot.

A supervisory management conference in York set up a counselling and information programme. There was a positive momentum in all this and we followed it up by writing to all the employees at home, making it clear in the letter that the union action was politically motivated and it would not help us in our negotiations on privatisation, during which we were working to protect their interests. The counselling, the letter, and most of all the good sense of our people won the day. The ballot was negative. In fairness, I believe this was the outcome Jimmy wanted: he understood the issues better than all of his executive.

The industrial relations struggles took the headlines, but the initiatives within the corporation were the mind-changers. These proved to employees that the management had a positive agenda which was in their interest and they were part of. As the safety statistics improved, so did workforce morale. Also, matching resources to need not only got the numbers and costs down, but it made each person a contributor.

*   *   *   *   *

Heathrow was complete and we had opened the Gatwick and Stansted dedicated lines, so our airport connections were secure. The skilled workforce available from the London-to-the-Tunnel project made construction of the Heathrow Express relatively simple to organize, rather to MacGregor's consternation. He had insisted on a common damages clause in which the party responsible for delays would pay. We were on time—and he was not! Ultimately it was cheaper for him to buy us out of the project, which brought us much-needed cash.

We wanted to go one step further and launch East–West Crossrail which would take passengers from Heathrow to the City in 38 minutes.

The treasury was uncooperative; although we were delivering results we were already eating into their budget. When we got it in front of the select committee at last and had Opposition support we thought we were home and dry. But John Major had not done his homework and his representatives voted it down. The attraction of 35 minutes from Heathrow to the City was not enough and, now well into the next century, it is still the Missing Link!

*   *   *   *   *

MacGregor produced a White Paper, *New Opportunities for the Railways*, which proposed splitting British Rail into two organizations: a track authority called Railtrack and a train-operating company. Railtrack would provide the infrastructure and be responsible for planning and operating the national timetable. The operating company would provide services until they were taken over by private sector companies under franchise. A regulator would license companies wanting to operate groups of passenger services, such as Network SouthEast or Scotland Passenger Services, and would pay a subsidy where necessary. Freight and parcel services were to be sold and maintenance was to be contracted from British Rail, which would keep responsibility for mega projects.

We were cautious in our reaction. We welcomed the clarity about the future but expressed concern about the complexity of regulation and the scale of bureaucracy. We stressed the need for investment and competent management and the fact that public subsidy was essential to underpin the continuing improvements. Robert Adley, chairman of the Commons transport select committee and a well-known railway enthusiast, was more outspoken—he described privatisation as a 'Poll Tax on wheels'.

The press were unimpressed. *The Daily Telegraph* saw it as taking 'five years to produce at best half a White Paper'. *The Independent* saw it as neither 'intellectually elegant nor administratively coherent'. *The Guardian* noted its praise of British Rail as being one of the most efficient and least-subsidised in Europe. *The Times* saw the supervisory structure of regulator and franchising authority as 'one of the most remarkable victories for Whitehall's bureaucratic centralism over market forces'.

In our document to staff, *The Way Forward*, we committed to keeping the railways running. Safe operation was paramount, and had to guide our transition to the new arrangements required by our shareholder. I stressed that the changes should take place at a pace and rate that everyone understood and could handle.

Adley pressed me on whether I would resign. I said I had not taken up the job for five years only to quit halfway through. And would I say exactly what I thought, he asked, to which I replied that I have never done anything else, although at times it has got me into considerable trouble.

After all this publicity, MacGregor asked to speak to our 45 key managers, although in the event he just asked for comment. The tone was set by the first speaker who described the gains we had made, likening the journey to crossing the Niagara Falls on a tightrope. 'It has been tough but worthwhile,' he said, 'and now you're asking us to go backwards over the same tightrope on a bicycle, risking everything that has been gained.' In fairness to MacGregor, he noted that great progress had been made and graciously acknowledged that it was ongoing. He hoped the same spirit would take us forward to implement his proposals.

It was a sad occasion. Sad because it made no difference and sad because MacGregor could see he was dealing with essentially capable people, and while they were not against him—they were too sensible for that—they were not with him.

At around this time in an interview with Nick Clarke for *The World This Weekend*, I said nothing had changed since my statements to the select committee, but I made a plea for a pragmatic evolutionary approach. The interview was widely covered—*The Times* forecast I would be the subject of Tory abuse. The *Telegraph* reported irritation in the department, and *The Guardian* had me at personal odds with MacGregor. By now it was almost Christmas and the storm clouds drifted by.

*   *   *   *   *

My first meeting with MacGregor in the New Year was cordial, and concentrated on getting the job done. Bob Horton, a former CEO of BP (and coincidentally another St Andrews' alumnus), had been approached some months earlier to join my board, and had told me he was being lined up for chairmanship of Railtrack. I warned him to think carefully about it, stressing that this was rather different from the well-organized oil industry (even if his own departure from BP had not been elegant). However, I said, if he wanted to come he would be welcome, and he did.

For his chief executive, I proposed Peter Watson, but for some unknown reason the recruitment consultants had ruled this out. They chose John Edmonds, which surprised me, and Peter Holmes' remarks came back to me. I was concerned that they had chosen a strategist rather than a strong operator with line experience, but I hoped John's high intellect would carry him through a period of major change. To be fair, it did.

Bob was impressed with the thinking that had gone into preparation for privatisation and the launch of Railtrack. The selection and allocation of people to key jobs went smoothly. In April 1995 we held our first two-day conference involving chairmen and chief executives of British Rail, Railtrack and its zonal directors, and the profit-centre directors. This produced a document we all signed up to, a basis for working cooperatively to the same end. This is a critical step in achieving change in large organization, and in our case it worked well.

The division of activities and responsibilities was detailed but the integrity of the machinery that ran the railways stayed intact, leaving the change to be understood and implemented by the leaders. At no time was the safety and smooth-running of the railways at risk.

The introduction of the rail regulator, eminent QC John Swift, brought a new process to policy-making. First a consultation, then a policy paper. In his consultation document he expressed confidence that competition would provide efficient service and bring about benefits to consumers. This, he averred, is what privatisation is about. However, after further consultation he directed that there could be no competition for seven years—by which time he would no longer be regulator! Having ruled out competition, the natural regulator of a market, he had to write himself into the system, which he did, through his projected regulatory activity. The brave hopes expressed at that long MacGregor briefing had been set aside for seven years. Journalist Simon Jenkins described the pronouncements as the completion of nationalization. The secretary of state was now the chairman, with all the appointments in his gift. For me, it was time to go.

*　　*　　*　　*　　*

Looking back, there are some achievements that stand out as exceptional, and others that represent lessons learned. Putting the confidence back into the organization after Clapham was a major milestone. Building in competence-based qualifications structures was also fundamental, as was integrating supervisors into the management structure. We stabilized industrial relations, accessing the strong common sense of a long-established union movement. Losing only two days in five years was an enviable record, particularly when 30,000 people left the organization voluntarily, taking £600 million out of the costs.

Such results were possible only with the support of a strong executive team and a sympathetic board. The physical results in those five years were also significant, the Channel Tunnel connection and the Heathrow Express being just two examples of many modernization and renewal projects we completed. Sadly, we were unable to make any impression on the policymakers, who had prioritised implementation of the political agenda over improvement of performance and services.

My time with the railways had at least one fundamental element in common with my time at Shell, and that was that I liked the people, and I liked working with them. There was also a similar trajectory: in Nigeria was 30,000 barrels a day in 1960, up to 1.3 million barrels a day in 1967—and then the civil war and nil barrels a day. And now the railways, as I left in 1995, broken up into bits. Alex Ferry, the outstanding union leader, commiserated. The revolutionary changes and great progress we were making had been waylaid if not stopped in their tracks, leaving the future of a much-improved organization to the whims and fancies of politicians.

Is all life like this, I wondered: you reach the summit then the black clouds gather and you realise you haven't made it at all? Well, if it is, I thought, you'd better dust yourself off and start again, which is what I did, in a whole different world.

10

London Electricity:
Keeping the Lights On

Sir John Wilson, chairman of London Electricity, was on the point of retirement when—having no doubt consulted government and gained his board's approval—he invited me to be his successor.

I met the team, whose knowledge of the network was impressive. This was similar to a Shell market where individuals bought their energy at the service station: here they bought energy from an organization that supplied their houses. The industrial and commercial consumers had direct supplies, again like the oil industry, and settled those on contractual terms.

Privatisation had moved the organization from the protection of government ownership on to the public stage where they had to answer to their shareholders. In many cases these were their customers. This made the AGM more of a public complaints hearing than an opportunity for the investor to learn about performance and future projects, but fortunately, the customer relations were excellent and had an answer for every question. The remuneration report produced a lively reaction, as did the investment programme, where there was a strong interest in what we were spending our money on and how much of it was for improvement of the service.

I had not been there long before new regulations required that all the organization's individual accounts be changed to a new format within a single accounting period. This was a daunting task. The format was complicated and the timescale almost impossible, but because the accounts produced the cash flow that supported the business, it had to be done. My auditors agreed to take on the project.

After six weeks the head of the firm asked me to lunch, and after we had finished our first course—a delay we could ill afford!—he gave me the bad news. They could not meet the deadline and were withdrawing from the project. However, he informed me magnanimously, they would not charge us. They offered no reparation for the time we had lost; it was a matter of principle across the sector, he said, that auditors did not pay their clients. This would be bad practice. At which point I thanked him and offered to pay for lunch. This he thought very amusing and I am not sure the sarcasm hit home—but as I left I wondered how he was going to get on auditing cryptocurrencies!

Meanwhile, one of the charities I was involved in had got itself into an administrative mess. Fortunately one of our supporters, an Indian company, offered to help and they quickly fixed the problem. Impressed by their efficiency, I asked to see the executive involved, Amur Swaminathan Lakshminarayanan (known as Lakshmi). I thanked him and asked if he would be interested in having a look at another piece of work that my company, not the charity, was involved in. I told him about the problem, emphasizing that the job had to be completed within the month once they had started taking the existing system to pieces. His eyes lit up and he said, 'I shall return within two days with a proposal.'

He did. His parent organization, based in India, had a large working resource in the United States. Using international time differences, Lakshmi could lengthen the working day from eight to 24 hours. He took on the job and completed it without a hitch, well within the time limit— and without my having to buy him lunch! This was my introduction to Tata Consulting Services, a member of the Tata Group, about which I had already heard good things.

*   *   *   *   *

The London Electricity team were frustrated by the absence of an attractive aggressive investment programme, so we stepped up the search until a North American opportunity surfaced. With London's increasing cash flow, the financials made sense as a basis for an outstep investment, and a proposal emerged. The economics looked good, but there were practical realities to consider, such as what did we know about the North American market and their performance in it? And how conversant were we with their regulatory environment? Our answers were doubtful, so we put the main issue back on the table. We had a company that was performing well but in the directors' view was significantly undervalued, and this was concerning our shareholders. Meanwhile, our financial advisers had identified interested parties who were prepared to buy us out at well above the market price. So we had a dilemma. Whose interests should prevail? Those of the management, who could continue to run a successful company and be rewarded for it, or those of the shareholder, who could monetize their investments? In the end the shareholders made the decision with their votes, clarifying the position for the board, their representatives. To the market we would go.

The most attractive offer emerged from Entergy, a US utility firm. We made contact and they summoned their oldest director to meet us, a cotton farmer who held great sway with the board and whose approval was essential. Jo and I spent an enlightening evening with him and his wife. At the outset he explained, 'I only get called in to make sure they aren't going to do anything stupid!' He admitted that that sounded arrogant, but where there was a possibility of expensive risk-taking, the opinion of a seasoned hand was always worth having. He went on to express confidence in us and the strength of the business. He asked if I would stay on the board, at least during the settling-in period, acknowledging that they had never done anything like this before, 'and while you know your staff, you never know how they will handle a new situation'. I agreed to stay, and with that, the deal was done. We returned to a discussion of Sweet Briar College, where Jo had spent a year thanks to the foresight of Senator Fulbright, and to the history of the American South and the making of the United States. It was a fascinating, enjoyable and profitable evening.

The UK's political environment was strange to Entergy, but they quickly got to grips with the regulations. The appointed chairman, a professional, experienced executive, settled in quickly. After 18 months I was confident that the company was running effectively, although privatisation in a conservative and structured industry inevitably disturbs the carefully designed human-resource planning. If individuals lose confidence in the stability of their position, they begin to determine their own paths, and competing organizations will naturally take advantage of that. At London Electricity, this led to a change of chief executive.

I found the new CEO to be outstanding, not necessarily in executive effectiveness, because that takes years to assess, but in golfing, which takes only 18 holes. As he warmed to his professional task, he gained the confidence of his management team, thus allowing me to leave with peace of mind. I had done as the old farmer had asked and it was now for his board to decide what to do with their investment.

As it turned out they were more comfortable with their domestic roots than with running a business across different political and financial regimes, which requires a certain type of executive and a flexible, experienced board structure. So it was not a surprise when Entergy sold to Électricité de France (EDF) and, in highly professional hands, the lights stayed on! For me, I had enjoyed my engagement with a successful industry.

11

British–Borneo:
a New Beginning

In the middle of 1993, a year into my involvement with London Electricity, Sir Douglas Morpeth invited me to dinner to discuss something he thought might interest me.

Douglas, born in Perth and educated at George Watson's School in Edinburgh, had served with the Royal Artillery in India, Burma and Malaya before leaving the army in 1947 to go to Edinburgh University, where he read Business Studies. On discovering that there were no credits for military service in the Scottish chartered accountants' institute, he qualified in England and joined Touche Ross (later to become Deloitte). He became senior partner and president of the Institute of Chartered Accountants in England and Wales, his work leading to improvements and reforms of the national tax system.

Having later chaired Clerical Medical, Douglas was appointed chairman of British-Borneo, which had been incorporated in 1912 as a limited company in the UK. By 1913 its interests included two exploration concessions in Brunei and one of 30,000 square miles in neighbouring North Borneo. In 1922 British-Borneo had sold all its concessions in Brunei to Shell in exchange for a royalty interest over any oil that might be found in the country. This was where, in 1956, I had begun my Shell career.

In 1926, the giant Seria oilfield was found. The royalty income from this field flowed into the coffers of British-Borneo which, having either relinquished or sold its activities in Columbia, Romania, Germany and Mexico, changed almost overnight from being an oil and gas exploration company into a financial investment company.

But for Douglas, this was not where the excitement was, so he cashed in his chips and set out to turn British-Borneo back into an exploration company. I was not entirely sure why he did this—was it to reverse the frustrations of the 1920s when the company had withdrawn, only to see others find huge accumulations of hydrocarbon and amass great wealth? Perhaps he did not realise the secret of the oil man is that he finds oil and makes money only to spend it again in more remote and more difficult areas! We are, I suppose, a special breed: the hunt and the chase are so much more important than the kill.

In 1989, then, many years since British-Borneo had had any full-time employees, Douglas hired Alan Gaynor as chief executive. Alan was a former BP executive and had formed his own company, GB Petroleum, which he sold on joining British-Borneo. His mandate was to liquidate the company's £20 million investments in a range of selected conservative assets, and to relocate the funds in exploration assets. It was back to the real oil business—if not in the jungle, then in an equally challenging environment: offshore in the North Sea. Douglas was right: it did interest me, and at his invitation I took over as chairman.

With an engineering background and a voracious appetite for work, Alan brought Douglas's vision to life. He inspired his teams not simply by force of personality, but by getting people truly involved. His groups were lively and filled with ideas, which he enthusiastically took on board and pushed forward. Alan had selected a first-class young finance director, Bill Colvin, whose hard common sense tempered his colleagues' wild enthusiasms. It was a good balance.

By 1993 the company's focus and objectives were clear and it had built up the competences necessary for active offshore exploration. British-Borneo was not just an investor in the North Sea, it was now an exploration company, unrestricted in geographical reach. Investments were securitized, returned or sold, and high-potential holdings in the Atlantic margin were held in a range of partnerships. Producing fields were kept for cash flow to support activity in the retained and new areas.

In the first seven years of his leadership, Alan not only strengthened his UK base but also acquired operational leases in the coastal offshore areas of Louisiana and Texas. By the end of the '90s, the company had an interest in more than 200 coastal leases, over 40 per cent of which were in water deeper than 650 feet.

From the perspectives of our coastal experience and successful drilling and production operation, the deep-water prospects in the Gulf of

Mexico looked tempting, if financially challenging. Also, offshore Texas and Louisiana were truly deep water: up to 4,000 feet—a far cry from our experience to date. But we pressed on, and the first project was on its way to production in September 1997. Alan had concentrated his engineering expertise on The Sir Douglas Morpeth Tension Leg Platform (TLP), and by standardising platform design and streamlining the process of construction, he had significantly reduced not only the delivery time, but also the cost.

Other opportunities began to emerge in the areas where we were active. Conoco, which had a 56 per cent interest with Shell as partner in a gas field called King Kong, offered its asset for sale. We discovered that two wells had already been drilled at the site, which was 40 miles south-west of Morpeth in 3,800 feet of water, suggesting there was substantial gas potential. Shell had already committed to beginning production during the fourth quarter of 2000. This was too good an opportunity to miss, particularly as cash flow would begin with production, so we bought it. Next, 14 miles north of King Kong and 30 miles southwest of Morpeth in 3,200 feet of water, gas had been discovered in a field called Allegheny. Having acquired Mobil's 60 per cent share, British-Borneo became the operator and work began on another TLP.

The asset base was growing, but there were setbacks in deep water. Only one of the three deep-exploration wells was commercially viable, which was a significant blow in this, the company's development phase. British-Borneo was beginning to learn that the ups and downs of exploration required deep pockets and a supportive press.

Alan had attracted Steve Holliday, Exxon's top UK prospect, to join us to explore potential outside the Gulf of Mexico. Steve identified attractive offshore opportunities in Brazil that represented yet another financial and operational challenge. We were moving out of the minor and into the major zone, but so far had neither the cash flow nor the skills to handle an international oil business without major financial input. We had certainly built up an attractive property by the time Alan and Bill called a meeting in late 1999.

They had analysed our position and determined that public disclosure of our financial requirement to cover costs would undermine the share price, which in turn would make further financing more difficult, if not impossible. An individual asset sale, they believed, even if it were feasible, would undermine the share price further. They concluded that the only option was to seek a deal for the whole company, so the sad process began.

The company's achievements in a decade or so were outstanding, particularly in two major oil provinces, the UK and the US. The management teams in both countries were excellent, as was Steve's opportunity team. The opening in Brazil had real possibilities for a serious investor, as was subsequently confirmed, while the North Sea acreages were certainly marketable. Meanwhile, an interesting potential venture had also appeared in Pakistan, and another in Cuba. It was an attractive package of assets for any investor with patience and finance.

After a period of look-and-see, the Italian national company Eni came out best, and I set out to meet the chairman of their exploration and production business. It turned out to be a surprising encounter. Our pricing aspiration did not seem to be a problem, so business was soon settled and the discussion turned to more general matters. He told me that he had lived in Port Harcourt, Nigeria, at the same time as I had. At the Shell oil camp on the edge of town we had built a club with a swimming pool and tennis courts. Since he and his family lived in downtown Port Harcourt they would not normally have been entitled to access the facilities, but he had applied anyway, more in hope than expectation. Happily, Shell had responded by arranging full membership for them and this move, he said, had kept his family together.

We arranged a further meeting with some of his colleagues at his golf course, where we discussed the oil situation and price prospects. Eni clearly planned to pursue assets, and we talked about its investment in Kazakhstan and interests in West Africa where Ghana and Angola had been successful. In the important matter of golf, the chairman was a ferocious competitor off a generous handicap and I only just managed to finish square. Colleagues who had worked with him in Port Harcourt joined us for the party that followed, and conversation was punctuated with hilarious stories of how they had tried to come to terms with their new environment when they arrived in Nigeria. Language was the real problem: if English had been difficult, pidgin English (spoken by many in Nigeria) had been completely impenetrable.

The next day we had lunch with the group chairman and, the deal sealed, he headed off to Verona for an opera, the head of exploration to Rome for a football match, and I to London to give the news to Alan and his team.

*   *   *   *   *

With hindsight, the weaknesses that led to the sale of this high-potential, high-performing company are clear. We were operating in a period of low, if not depressed, oil prices. Investments had to be cash-positive at $16 a barrel and at the time this was hard, if not impossible, for a company in its infancy to achieve. In addition, one major part of the activity, moving into deep water, was new to us. Drilling deep wells and operating in the open sea are vastly different from working in coastal waters. Unlike our competitors, we had had no long period of apprenticeship and no opportunity to learn from our mistakes. Entering high-pressure reservoirs is a high-risk operation, as BP's tragic experience in the Gulf of Mexico would prove. Being a small player in a big boys' game is never comfortable. It was time to hand over and move on.

12

A Sale at Sears

The apparent rudderlessness of the Sears organization was giving London's banking fraternity increasing cause for concern. Since I had now established my credentials in the context of company sales, I found myself among those approached to help find a way forward. Once we had determined that there was no corporate body prepared to steer Seers' combined collection of assets, we decided that the best way to achieve shareholder value was to sell the constituent parts individually.

The electricity industry delivers in specifics to a large and diverse customer base; the oil industry delivers in bulk from a distant source. Although in both cases millions are spent creating the myth that one company's product is different from another's, differentiation is not an issue since the product is never seen—or oil only momentarily, as it disappears into the tank. People managing the movement of oil and electricity have to stick to procedure and process. There is no room for flair, and if there is something to see then there is usually cause for alarm. At Sears the goods for sale were on display; you could touch and feel them. The individuals involved had flair and imagination. Differentiation here was the very secret of success, and the market was made by endless diversification. There is nothing more refreshing and attractive in the world of fashion than change!

I needed to learn how the business of tangible product worked, so I started with British Shoe, which had a range of shops with established names such as Lilley & Skinner, Dolcis, Saxone and Shoe Express. There was no shortage of energy and activity at their offices and warehouses in Leicester, but the figures were not adding up. I came away with the dispiriting feeling that things were only going to get worse, in spite of the fact that a consultant had been appointed.

To find out what a good shoe business looked like, we sent our man out to the Far East where the shoes were made. The suppliers had agreed to show us how our performance compared with that of their best-organized customer. Our man came back with a graphic description of how the British shoes were bundled into containers and packed off to Leicester, where they were then unbundled, put into boxes and stored until they were called for or rota-delivered to outlets. An American competitor, by contrast, had their shoes packed in boxes at source, ready for presentation to the customer. What's more, the boxes were shrink-wrapped and addressed to the actual point of sale. All the handling costs were covered by the manufacturer so warehouse and security charges did not apply. There was no central store, no double handling: each pair went from Asia straight to the retailer. The next step was the customer and the cash.

This was the nub of the problem: our system was back in the '40s, and without a major transformation we simply could not compete. Even allowing for our marketing expertise (and some of it was top-notch), we could not begin to bridge the gap. We had other businesses to turn around and cash flow was haemorrhaging so, having reviewed the position carefully, CEO Liam Strong and I and decided that the shops should be sold and the structure that was British Shoe dismantled. As a natural part of this, the consultant was released!

In closing British Shoe I drew on my experience of downsizing Shell UK's Carrington Chemical plant. The consultants I had used there came up to Leicester to manage the process. 'Take the fear out change' was their mantra, and sympathetically and carefully they helped individuals into a new frame of mind, shifting from the dismal prospect of being 'let go' to the opportunity of having time to rebuild, retrain or do something else. It was money well spent; we owed that at least to the staff and to Leicester.

The shoe story with its £150 million accounting loss was seized on by the press and the attacks on Liam grew as quickly as his image as a golden boy had sprung up. He was a tough individual with business in his blood, having grown up in County Tyrone where his father ran a lorry fleet for the local milk marketing board. His humour was irrepressible but his confidence was waning. Against my advice, he accepted an interview at this point with journalist Patience Wheatcroft, whom he liked. I warned him there was no sentiment in the media; they were trained to go straight for the jugular if it suited them. When he returned he assured me that he had got our side of the story over … and then the article came out. The opening paragraphs seemed understanding but the conclusion was vicious and cruel. After this Liam had no stomach for the long haul, which by then we knew we were facing. It was time to tell the board.

With Liam gone the board executive took up the challenge and we worked together on a strategy for the rest of the group, our first objective being to protect shareholder value. We started with Selfridges.

Harry Gordon Selfridge was an early-twentieth-century business icon who from 2013 was reinvented as a television star. He was born in Wisconsin in 1858 and began his customer contact early by delivering newspapers from his father's store. Aged 14, he left school and joined a bank. His acumen was noted and he was recommended to Marshall Fields, a huge store in Chicago, where he began to build his experience and retailing expertise. Over 25 years he amassed a fortune and married a wealthy and able woman, property developer's daughter Rosalie Buckingham. The couple came to London in the early years of the new century and invested £400,000 in the then-unfashionable western end of Oxford Street. In 1909 he opened his store where customers were urged to shop for pleasure, not for necessity. The goods were on display and accessible, and staff there to assist the customer. It was an enormous success.

Sadly, Rosalie perished in the flu pandemic of 1918 and Harry Gordon's mother, who had always lived with them, died in 1924. His anchors gone, his life fell into disrepair, the decline accelerated by his free-spending habits. He remained chairman until 1941, but before he died in 1947 he was reputedly refused entry into his brilliant creation as an unwanted vagrant.

Philosophizing is dangerous but the story illustrates at least one undeniable lesson: success does not guarantee happiness. That comes through focus and concentration, but often at the expense of the other elements that make up life. People need balance, and friends who will tell them where and when they are losing it.

After Harry Gordon left, a Liverpool chain called Lewis's held ownership of Selfridges until 1965, when it was acquired by Charles Clore and added to the Sears Group.

*   *   *   *   *

We set about de-merging our asset. Harry Gordon's treasure was going to be on its own again. I selected as chairman the chief executive of Avis, Alun Cathcart, whom I had seen confidently handling both the board and the shareholders. To work with him we had Vittorio Radice, an imaginative retailer who had already revitalized the shop and was, in the spirit of the original merchandising movement, paying tribute to Selfridge himself.

After Harry Gordon's wife died, he had sought succour in music, dancing and theatre. Vittorio had photographs of dancers of the era and commissioned sculptor Edoardo Paolozzi to produce a work based on them. Edoardo sculpted the French entertainer and Resistance fighter Josephine Baker, whose pioneering spirit had something in common with Harry Gordon's own. Edoardo, born in Edinburgh of Italian parents, was intense and passionate, with massive hands, and shoulders like a bull's. The sculpture was beautiful, and I felt Harry would have loved it.

Vittorio and Alun quickly made the most of their new and expanded roles. As always, if you find the right people, give them their heads and are there to support them when they need it, personal chemistry will drive things forward. The flotation was a success and the strategy under way.

The next target was to sell the mail-order business Freemans. Researching another industry unfamiliar to me, I learned that the mail-order cycle began with new-product marketing. This was followed in time by sales and reductions that took the prices close to break-even, and the last stage was sending the goods in containers to the Eastern European market where cost-recovery was the target. Freemans was a well-managed business with a good record and sound financials. However, as with British Shoe, I wanted to learn what a first-class mail-order business looked like and how we stacked up against it. I chose N Brown, and found that one significant difference between the two businesses lay in their respective telephone-sales technique. At N Brown's, revenues rose per call as a result of the staff's outstanding marketing skills. It was interesting to see in practice the principle Harry Gordon understood so well: customers do not really know what they want until the alternatives are put before them. This gives the marketer the opportunity to make the sale.

When N Brown's owner knew I was looking to sell, he declared his interest and came with an associate to see me. I provided the financial and operational data and, confirming confidentiality, he left with his chum, promising to return. When he came back, I told him my price, and after a bit of haggling we came to an agreement and shook on the deal. Our lawyers would meet and put it in legal form.

After a few days he was back with his friend saying that the price was too high and he wanted a reduction. Amazed that he was not now prepared to stand by his agreement, I reminded him that we had had a witness to it. He started to protest but I stood up and told him that if he wanted the business, that was the price; it was not negotiable. He left with his chum, who had not said a word. I had a feeling the man had seen this happen before, when perhaps the ruse had been successful—but not this time!

I turned my attention to Littlewoods in Liverpool. James Ross, whom I knew from BP, ran this company and was willing to talk. After a few sessions we established that he was interested at the price we had originally agreed with N Brown. This looked like a realistic proposition and I was confident it could be completed.

The rest of the portfolio was mostly women's fashion. Wallis was well managed; sold good-quality product with the emphasis on style rather than price, and through skilled marketing had built up an enviable set of clients. This would be an easy disposal and one in which we could be confident that the staff would be looked after by the new owners, since they were the business.

Miss Selfridge, like Wallis, was an excellent business; there would be many potential buyers for these treasures. The merchandise was not of the same quality, but it did have the name, and that alone would generate great interest.

At Warehouse, brand director Yasmin Yusuf was a fashion leader for young and/or young-at-heart women. She had been trained in retail fashion and had flair and imagination. Her exciting ideas attracted young people to join her, not just for a trip to Malibu for design inspiration, but also because she was creating a go-ahead business. On my first visit to her office I wanted to learn about pricing. The team was mostly young but there was one older woman sitting quietly at the back of the room. Although it was August, the discussion centred on a warm winter coat, attractive photographs of which were already appearing in their outlets. It would be in shops at the end of September and would be a sell-out, was the message, and the list of interested customers was growing. I asked how they would price it. As I understood it the formula was usually cost of manufacture plus mark-up, then comparison with competitors, then trial and error ... but these ideas were quickly thrown out. Eventually the older woman, who was French, spoke. 'Pricing is not a science; it's an art,' she said. 'It's an art that needs a long apprenticeship to perfect. Knowledge of the customer is an important as the article itself.' She then put a price on the coat and everyone nodded, except the boss who said,   'We'll be lucky!' But Madame was right. They were lucky, and the coats sold out by early autumn.

Reviewing our overall position, I felt we were implementing a sensible strategy. The shoe business, our biggest loser, had gone; Selfridges had been successfully de-merged; we already had a good price in prospect for Freemans, and the retail properties should sell well with the exception of one asset we still had to work on. Adams Kids was clothing for the children's market. Its performance was solid but I was concerned about its proposed strategy to move out of its comfort zone to reach wealthier customers. The company could lose its established market if it pushed too far, and might not be able to build a higher-price market. It did not have the in-depth skills, I felt, to make this a successful venture, and its locations were also against it.

*   *   *   *   *

At around this time we were about to take a mysterious item on the agenda at a Bank of Scotland board meeting when I was asked to leave the room. A number of possible explanations came to mind and admittedly Sears was one of them. When I re-joined the board I was informed that the Barclay brothers, two well established entrepreneurs, had joined forces with Philip Green to bid for Sears. I had no negative feelings about the brothers' involvement; it was a natural business opportunity that had come to the bank from an established customer. Philip Green I did not know so could not comment on.

The Sears board was immediately informed. Our financial advisers were clear: Mercury Asset Management held close to 40 per cent of Sears' shares and would back the bid. This made it an easy decision—to fight and lose would be expensive and would jeopardise the end price; we would end up paying twice. We made a straightforward calculation of what we could get for the retail businesses and the mail-order investment, and set a minimum acceptable price. If they were not prepared to pay that, we would think again.

A meeting was set. The aggressor chose and paid for the venue—a huge suite with anterooms at The Dorchester. The Barclay brothers had sent two of their sons with Philip Green. We started off with their price to which I said 'No', remarking that I had understood they were serious bidders. Philip Green let off a diatribe, casting aspersions on all and sundry, at the end of which I enquired, 'So your proposition does not see you taking on any of the management, then?' Philip laughed, said definitely not, and went back to his tirade, at which point the elder Barclay son asked him to proceed forthwith to the anteroom. I was impressed with their foresight in choosing a suite with anterooms. Perhaps they had had dealings with Philip before.

In a calmer atmosphere I laid out my sticking point and the logic for it. I said I had the mandate to fight and would be happy to do so, although their fathers might not be too pleased with that outcome. Reluctantly they agreed and Philip returned. After indignant protest that they had been robbed, he took his orders from the financiers. I let them know that I had a deal in my pocket for Freemans, but Philip swore he would never accept the price. I learned later that they eventually sold it for significantly less. I returned to the management issue and Philip reiterated that he did not want to keep them on. The sons acceded. We shook hands and as I left I asked the Barclays to give my best wishes to their fathers, for whom I had great respect. Their charitable gifts to a teacher training college in the East End kept that institution alive and did much good work in local schools. They were genuine philanthropists.

Back at the office I told the board that the deal was done. I then saw each member of the management team individually and advised them of their options regarding redundancy. The finance director opted to stay but I asked the personnel team to make the calculations and pay the others before the end of the day. I shared with them my suspicion that they would be invited back, and that if they were they would be able to set their own terms. Their finances were protected and they could start their new—or rather old—assignments on a fair negotiating basis.

Two days later the negotiating team arrived in my office. They too asked to meet the management team individually and politely said they did not want me there. This would take a few hours, they explained. 'No, it will take one minute,' I replied. 'With the exception of the finance director they all accepted that you did not want them.' I added that I would supply all necessary contact details just in case they had second thoughts. In fact the team members were all rehired and, as far as I could discover, mostly on more advantageous terms.

Winding up Sears had been a sad business, but it was done. The director who stayed with me throughout the process, offering consistent and continuous support was Norman Tebbit. I understand why Thatcher regarded him so highly.

So Sears, like Harry Gordon, passed into the history books. For the others on the team all went well: one ended up a knight with a Chevalier de l'Ordre National du Mérite from France and a multinational business to look after, another tangled with the secret work of MI5 and another ran his own large womenswear business.

Dismantling a conglomerate in the retail and clothing business and accessing real value for the shareholder had been a challenge, but in the end I was satisfied that as far as possible we had achieved it, while rewarding the shareholder and safeguarding the employees.

13

Bank of Scotland

The Bank of Scotland, which opened for business in 1806, holds a commanding position on The Mound in Edinburgh, its rear windows overlooking the New Town down to Princes Street Gardens. The Presbyterian assemblies are held a stone's throw away, and further west lies the castle—a valuable repository for cash during revolutions. The bank's front door faces the Old Town and its previous quarters in The Cowgate.

The staircase leading up to the bank's boardroom is lined with paintings of past governors; serious and substantial portraits of men who clearly knew the gravity of their task.

When the time came for the new governor, Sir John (Jack) Shaw, to sit for his portrait, he insisted I join him as we had shared the deputy governorship role and served together as non-executive directors. Jack selected Romanian-born French–Israeli painter Avigdor Arikha as the artist and we sat, or rather stood, individually in Arikha's Paris studio for the portrait. It was my turn to be painted on the second day, and after three hours' work the artist exclaimed, 'I have you!', inviting me to take a look. To my amazement all he had painted were my eyes. And so to lunch, and then another session before we eventually headed home.

When the finished painting arrived, it was as different as it could be from those august portraits on the staircase. Jack and I were standing apart from each other, looking in different directions into the far distance. Both faces were rather expressionless. In the background on my side was the sea, dotted with rigs. It brought home to me how odd I looked with one hand, my empty sleeve dangling in the wind. Was this symbolic? I didn't think so. But perhaps Arikha felt that Jack and I were looking for different things—and that the man with one hand needed a hand!

After a successful spell in a New York art exhibition, the portrait finally turned up at The Mound. Where to put it? The staircase was obviously out: the image of this untrammelled pair did not sit easily alongside the earnest men at their sombre desks. After some delay it appeared in a small meeting room on the ground floor. The reception staff confided that the room had never been as busy as it then became, so Arikha could count that, at least, as an achievement.

The historic images inspire one to look further into the bank's origins. William Paterson (whose only remaining portrait hangs in the British Museum) was born in 1658 in southern Scotland. As a young man he moved to Bristol, and from there to the West Indies, where he saw at first hand the promise of a New World and the rewards of merchant adventuring. It formed in his mind a dream which in time, sadly, would be his undoing.

When he returned to England, the country was at war with France. Wars are expensive affairs and the king was short of money for his war chest. Paterson saw this as an opportunity, and with the support of wealthy merchants founded a bank to lend to the king and the government. In a relatively short time the royal charter was given and the Bank of England was created—by a Scotsman—in 1695. What was significant for the future of this new financial industry was the undertaking to raise over a million pounds. It was raised not in hard money but in bills of exchange, which were commercial promises to pay. These ephemeral promises earned 8 per cent annual interest from the government, effectively making money on money that did not exist. As Roy Perman says in his book *Hubris: How HBOS Wrecked the Best Bank in Britain*, the alchemy had begun: wealth was being created out of promises, building castles in the air in the expectation that the underlying pledge will be fulfilled.

Paterson, a restless and ambitious man, soon fell out with his fellow directors and headed back to Scotland, leaving England with not just a bank but, as it turned out, The Bank. His new vision was even more ambitious. He had seen the potential of the Americas and the success of the East India Company, and believed it could be repeated in the West, so he directed his energies to creating an East India Company for Scotland. He chose Darien, on the Isthmus of Panama, for his new settlement. Paterson dreamed of a trade route to join the Atlantic with the Pacific, a potential

Panama Canal—but the venture ended in disastrous failure. Disease and frequent hostile attacks left hundreds dead, including his wife and child. The calamity was a national one, financial losses affecting the thousands of Scots who had flocked to subscribe. It was written into the history books and etched in the nation's psyche. I remember reading about it as a small boy and wondering why Paterson did not foresee the catastrophe. It was a question I carried with me throughout my adventures: Am I entering a disaster area? In a life filled with fast trains, helicopters and offshore platforms—not to mention blowpipes, head-hunters and the odd airborne hog—it seemed a reasonable enquiry.

At the end of the seventeenth century and just before the Darien scheme was launched, a consortium of Scottish businessmen proposed to follow England's example and form a Bank of Scotland. They chose John Holland, not William Paterson, as their chief executive. Holland was English and his CV read very differently from Paterson's. His commercial education and experience were impeccable and his visions and ambitions soundly based. He had gained proficiency in accountancy in Holland and then returned to London to become involved in financial affairs.

Unlike Paterson, Holland and his board were apprehensive about lending to the state—indeed, the bank's charter prevented it from doing so. It was designed to finance the private sector and make its fortune in the commercial world. The controls the prudent Holland put in place required that any agreement to lend was examined by committee, rather than by just one person, and the terms of the loans were set to protect the bank's liquidity. By the time I joined almost 200 years later, these standards were enshrined in the bank's ethos. They underpinned not only the confidence of the executive, but also the bank's financial strength.

*   *   *   *   *

As the last decade of the twentieth century began, profits had again risen by 25 per cent and capital had been raised from shareholders to maintain the balance sheet. The question of how this growth was to be sustained and financed was already causing concern.

The governor, Bruce Pattullo, realised the importance of standards and clarity of thought when facing the challenge of growth. Throughout his years of executive leadership he had imposed an intellectual discipline that monitored and controlled lending behaviour. He instructed me to rein in my aggressive acquisitive instincts, and as the individual loans came up for homologation—endorsement by committee—he taught me that the bank was interested only in the safety of its money and the return of it, not in the value of the borrowers' shares. Naturally there was a relationship, but clear criteria forced the executive to examine the cash flow and its accessibility as bankers, not as long-term investors. My position to that date had been that of an investor, so it was duly adjusted!

Homologation reminded both board and management that in the real world of human assessment it is not difficult for ambition to outrun common sense, and that a second look at the detail was invaluable. For management, this was a testing process and one in which board members often knew more about the borrowing corporation's standing and activities than the bank executive did. The loan application might look good on paper but there were other things—potential court cases, land disputes, internal strife and disagreements—that were not always disclosed during negotiations. I could often contribute to the process because of Shell's extraordinarily wide circle of contracts.

As my discussions with Bruce continued in the boardroom and outside, I began to realise the significant changes that he and Tom Risk had wrought during their time at the top. In an interview with the *Financial Times* when he was appointed group chief executive, Bruce declared his intention to make the bank 'the best-performing bank anywhere'. He was already making progress towards that goal. In the 20 years prior to his appointment he had seen outdated practices and organizational structures hold things back—now he had the chance to streamline his inheritance. This he did, and at the same time opened up opportunities for energetic new executives, revitalizing the bank and its ambitions. He sought and seized commercial opportunities, oil and the North Sea being primary targets, and the bank's name and cash became associated with a bonanza of activity.

Restrictions on trading in England had arisen from the belief that you should keep your customers within close range. These restrictions were now set aside and success stories grew. The bank under Bruce Pattullo knew and understood its customers' activities, and its mandate was broadened only with the utmost care. Understanding his team's limitations, Pattullo selected and purchased established performers in particular fields rather than attack the markets with his own people who had little experience in them. He bought the expertise of Chester-based North West Securities (later Capital Bank), a firm whose consumer credit business grew under a team of highly effective, adaptable executives. Imaginative deals with retailing giants such as Marks & Spencer and C&A were complemented by the innovative introduction of the Home and Office Banking System (HOBS). This was the precursor of internet banking and brought in a large number of accounts. The bank at that point was alive and well in all sectors, and its appetite for expansion was growing.

Early in my relationship with the bank I was involved in considering a paper, a valuable insight into the bank's strengths and weaknesses, by Robin Browning. It began by setting out the underlying capital structure: the Basel ratios. Capital supporting the bank was defined as Tier 1 and Tier 2. Tier 1 consisted of the ordinary share capital, reserves and irredeemable preference shares. To increase Tier 1 would require a new capital issue, a reduction in dividend, increased profits and other less straightforward solutions such as capitalising property revaluations or even reducing bad debt provisions, none of which seemed remotely attractive. Nor did the sale of the carefully accumulated assets. A new issue of preference capital, however, seemed a sensible option.

Tier 2 capital was divided into upper and lower segments, the upper containing undated loan capital, general provisions and property revaluation reserves. The lower contained dated loan capital, which had diminished over the last five years.

Taking Tier 1 and Tier 2 together, the bank had £1.5 billion in 1990. This was the basis against which its risk-weighted assets would be measured. Cash carried no weighting, mortgages 20 per cent and commercial lending 100 per cent. The more you lent commercially, the faster your risk-weighted assets grew and the capital base, if it was not augmented, decreased as a percentage. Acquiring mortgage business was a more attractive alternative, and this could be effected by taking on a building society (which in turn would optimise the building society's deposit base). Each element had to be watched and carefully balanced to safeguard the bank's viability.

Browning's paper pointed out that the Tier 1 ratios over the last two years had slipped to 5.1 per cent against the Basel minimum of 4. Tier 2 had also reduced to 9.1 per cent against Basel's 8. In addition, Basel convergence required that Tier 2 capital must not exceed Tier 1 capital. With the balance sheet growing at well over 10 per cent, our risk-weighted asset base was putting increasing pressure on our ratios. This had come about due to our success: our assets had doubled in under five years and growth was currently running at over 30 per cent a year.

Browning projected that at the current rate of growth, assuming no new issues of Tier 2 capital, the Basel minimum would be breached within two years. Even if the Tier 2 capital were raised to the maximum allowed, the target ratios would be breached in three years.

It was salutary to see that improvement would come faster from reducing asset growth than from increasing the return on investment. In fact, the comparison showed that the beneficial effect of taking the foot off the accelerator was twice that of doing more profitable business. This hard reality was aimed at the gung-ho management and, as the author intended, it hit its mark!

The reactions that were then agreed are interesting, considering what was soon to transpire. Increasing Tier 1 and Tier 2 capital was an obvious move. Capitalising our property revaluations was also sensible, but the reduction in provision of bad debts less so. There was much discussion and focus on the growth of low-margin lending. Browning's scenario had called for new retail deposit products, which implied that we were missing opportunities.

It was a seminal paper and as a non-executive board member I was pleased to see that the bank had brakes as well as an accelerator. As long as this remained the case, the lenders' animal spirits would be reined in. There was considerable tension between them and those who held to the banking testament *ne quid nimis*—nothing in excess.

In May 1996, before these concerns could be fully resolved, the unexpected happened. Having taken a long hard look at its portfolio, Standard Life's fund managers decided their shareholding in the bank was disproportionately high. They planned to divest and place the resultant cash elsewhere, which would both reduce their exposure and spread their risk. At least this was the story at the time; I wondered then, and with hindsight I wonder even more now. Perhaps the taciturn gentleman from Standard Life who sat on the bank's board could see the clouds on the horizon. Pattullo would soon retire, leaving the unresolved funding issues in less reliable and less risk-averse hands. Perhaps he felt it was a good idea to go out on a high, and after all it was a market so he could always come back if he chose to. As it turned out it was not difficult to place their holdings: the bank had many potential supporters and they quickly came forward.

In spite of this diversion we concentrated on the primary problem, the increasing dependence on wholesale funding that was weakening our capital base. We urgently needed to find a new source of deposits. The obvious answer was to acquire a building society, but these were staid, settled organizations that had no appetite for movement. Discussions with them were mostly polite, although some institutions seemed annoyed by the apparently arrogant suggestion that we could make better use of their deposits than they could. Some were aggressive in their refusal to discuss, some were delighted with the social contact ... but none was tempted.

Sir Bruce duly retired in 1998, admitting defeat on the building society front, and Sir Alistair Grant took over, Peter Burt remaining as chief executive. We continued to seek the ideal partner until early in 1999 when the possibility of a link up with the New Foundation Bank in America emerged. The US connection was with a Dr Pat Robertson, who had built a television channel with 55 million subscribers, all potential customers for financial services, and all naturally open to promotions. The mechanics of this deal were attractive and a target of $700 million deposits in year one, rising to $3 billion in year three, seemed feasible. There was just one snag.

Robertson was a regular performer on his own network. His outspoken condemnation of homosexuality and his views on other religions created active opposition. The controversy grew to the extent that it could seriously damage the bank, and as more protest groups appeared and added fuel to the flames, Robertson joined the fray. This was the final straw. The deal was cancelled.

Peter handled the ensuing public-relations disaster well, but the venture had been an avoidable mistake. Pattullo had warned against wanting something so much that you ignore your better judgment; a modicum of research would have exposed the dangers.

The business continued to be very profitable, with a return on equity running at 22 per cent after tax. The Royal was the only bank doing better than this. Our cost-to-income ratio was the best in the industry and by all other measures we were well up with the pack. On the back of this we confidently set out to generate a substantial increase in dividends by running our business more effectively and developing new markets. We had had some real success: auto leasing, telephone banking and finance for local motor dealerships were good examples. Lending to businesses that had arisen from the government's private finance initiative was another new and exciting source, and there were others. As we moved through the last decade of the century, I was confident we were well set on a fair wind.

We had expanded significantly via a number of acquisitions—the Bank of Wales, Countrywide Bank (New Zealand), and others—so we set ourselves five criteria on which to base the pursuit of more. First, no acquisition should endanger the bank's viability and independence;

second, we should not undertake anything that had both a geographic and a functional diversification; third, the target must have an actual or potential attractive market share; fourth, we must be able to add value to the target, and fifth, the transaction must make economic sense—in simple terms, we must not pay too much.

Industry analysis, meanwhile, had identified weakness among the majors. Nat West was a loss-maker as we saw it, and over the last three years their return on equity was well behind ours and the Royal's. Their cost-to-income ratio was in excess of ours and their asset growth was falling behind. It was not long before we took a serious look at it as a potential acquisition.

In any stock exchange transaction you are surrounded by advisers, none of whom can guarantee success. Our selection of advisers for the possible Nat West deal was no exception. For them, there is no penalty for failure—not even a reduction in fees, so it's a win-win game!

We looked first at the level of cost-savings the combined company could achieve, which we calculated as over a £1 billion within three years. Then at the revenue potential unrealised by a sluggish Nat West, which was primarily concentrated in personal banking. The marketplace also was well aware of Nat West's decline. Profits had been below expectation and towards the end of 1998 it had to announce a £350-million exposure to Russia and £75 million of additional trading losses. By the end of November, Derek Wanless, the chief executive, had resigned. In September 1999, the Bank of Scotland made its offer at a premium of 21 per cent to the Nat West share price.

Nat West's reaction was disappointing. Each of their criticisms of our proposal could be rebutted just by looking at their record, and as the days passed and the likelihood of an intervention by the Royal Bank grew, they made some odd proposals. The possibility of our chief executive joining them to negotiate a merger from inside was one of the strangest, but a clear signal, nevertheless, that they were caving in and their future would soon be in someone else's hands. But whose?

By February 2000, the Royal Bank had stepped up the pace providing figures that purported to show that its bid offered a significant value advantage over the Bank of Scotland's. The press and financial analysts judged that there was no clear winner, and were still not ruling out the possibility that Nat West could escape, which is what its new executive chairman, David Rowland, was trying to achieve. I like to think that had he been in favour we could have staved off the Royal at the last moment, but it was not to be. The Royal won, and Fred 'the Shred' Goodwin took all the advantage and more that our earlier analysis had revealed. But success can lead to hubris, and hubris drives in only one direction.

Having missed the prize we had worked so hard for, we had to pick ourselves up and start looking again. Sticking with the status quo was no longer an option. 'Scale', we felt, was critical to survival, so we had to buy it or build it. The protagonists of the 'survival' view gave us only ten years before we would be swallowed up by one of our competitors, and in our present state ten years was not enough time to secure our future. Looking back at what happened over that time, I believe the protagonists were mistaken.

The *1999 Report and Accounts* gives a clear picture, and we should have studied it more carefully before rowing on. In that year the clearing bank had a pre-tax profit of £480 million, 42 per cent up on the previous year. Customer lending was at £28 billion. Capital Bank was up 12 per cent at £165 million, with customer lending at £11.5 billion. Treasury services and our merchant bank activities were close to £75 million, with lending over £16 billion. Each of these businesses had a range of possibilities to achieve growth—the merchant bank had strong prospects with growth rates at well over 30 per cent. In short, if somebody came to take us over they would have had to pay a high price, and that would have assuaged the shareholders' pain.

But there was more to consider. Bank West in Australia was producing AS$183.7 million profit pre-tax, with customer lending at AS$13.4 billion and a growth rate of 12 per cent. Our international activities were successful and our New York office was achieving good commercial business and returns. This gave us a sturdy foundation to build on, so that might have been the better path, but we gave this alternative too-short shift.

We surveyed our market and made one or two tentative forays, but they served only to unsettle our hosts. At last we hit upon something that looked like a good fit. Halifax Building Society was capitalised at £16.4 billion, with total assets of £183 billion and net assets of £6.8 billion. They had 836 branches and 29,000 employees. Its board was chaired by Dennis Stevenson and the chief executive was James Crosby; the primary retail business leader was Andy Hornby, who had already made a name for himself at Asda. The strong team were proficient in business and well controlled by Chief Operating Officer Mike Ellis.

In our house, Jack Shaw had handed over to Alistair Grant, who had also made his name in the retail market with Safeway. He was extremely capable and had seen banking from the customer end. He would, I believed, make a first-class chairman in this new venture. But then tragedy struck. Alistair was diagnosed with cancer, although he was told he had a good chance of recovery. Then he was advised that the condition was terminal. In human affairs nothing is sadder than having all your prospects and the future disappear in front of you. In helicopters and oil platforms, in trains and railway lines the end is immediate; in a sickbed it is slow and inevitable, and you almost have to watch it happen.

And so it was we had Stevenson as chairman, Crosby as chief executive and Peter Burt as deputy chairman. Hornby ran the retail and George Mitchell the commercial banking. The announcement of the deal in mid-2001 confirmed these arrangements and went on to define Peter's role as responsible for overseeing the integration of the two organizations 'with a view to the most efficient and timely achievement of the substantial revenue and cost synergies […] establishing the organizational structure to ensure cross-selling of financial products and services between the two organizations'. It also named Mike Ellis as group finance director.

Looking back, I wonder whether these were the best roles for these two successful people. Did Peter have the power, from what was essentially a staff role not a live executive position, to do to Halifax what Fred Goodwin had done so well with Nat West? Could Mike regularly meet the stringent requirements set out in Browning's paper to bring the bank back to reality? History unfortunately gives us the answer as Crosby, Hornby and Stevenson were pilloried in both the press and in parliament, along with George Mitchell's successor, Peter Cummings.

*   *   *   *   *

Were there signs in 2004 when I left the bank that things were beginning to slide? There was certainly one incident that gave me cause for concern. Shortly after the merger the stock market fell through the floor—and it was not finding a new floor. At this point one of the executive directors accused the chief executive of contributing to the market's weakness by continuing to sell stocks into it to meet the regulator's liquidity requirements. The Bank of Scotland, he averred, did not buy into Halifax just so that the CEO could destroy the bank's assets. The meeting ended unhappily, and on my return to London I rang the regulator. She said it was obvious that the liquidity requirements should be suspended, but admitted she had not so far persuaded her management to act. Hearing this I rang Eddie George, governor of the Bank of England, who had been helpful in the past. He was sympathetic but explained that under the new arrangements it was the Financial Services Authority that held the power. I replied that that was where I had come from, and that this was a serious enough situation to override organizational division. With that he said he would do what he could, and within hours the avalanche of stock halted and equilibrium was restored. Eddie had acted.

I can now see that this was a significant moment, not only because neither the chief executive nor the chairman could find a solution, but also because the bankers had lost faith in their new partners.

At Shell I had benefited hugely from being carefully prepared for senior positions, and had suggested at the time of our merger that Andy Hornby spend a couple of years in Australia under an experienced chief executive learning about banking. The suggestion had not been taken up. In time the true bankers, who were by force of circumstance running the show, drifted away one by one. This left an unprepared executive to handle a banking crisis which they—unknowingly—were creating. The restraining hand of the intellectual banker had gone, and with it went the restraining hand of homologation. Had the process remained in place, it might have acted as a check on the rash lending of the last ten years. The bank's demise as an independent entity was a great sadness to me. It is little consolation that all over the world Scots hold and have held the highest positions in banking and finance.

Contemplating success is pleasant, the opposite is not, but responsibility rightly falls on the shoulders of the leaders and the business environment they created. The sad thing is that these were good people out of their depth. To examine where and when the seeds of the final disaster were being planted, one must examine what was going right and why we were confident the success would continue.

The first two governors I served with, Tom Risk and Bruce Pattullo, faithfully followed long-established processes: first the regular review by an executive outside the business's line of management that was designed to reveal the dangers and indicate where correction was necessary. It presented the facts and made its projections with unflinchingly honesty. After Pattullo left, this process was replaced by line-management reports which, one after another, glowed with achievement, failing to acknowledge any downside.

Secondly, the process of homologation was dropped, and with it the opportunity for non-executive directors to question the lending of the executive. It might seem an odd idea to examine decisions after they have been made, but it worked. With the death of Alistair Grant, the character of the bank changed. The management became building-society driven as the bankers moved on.

Before these losses, I had taken the opportunity to look at the new activity we had become involved with. The evening I spent with Andy Hornby and his team had been refreshing. Their enthusiasm and confidence should have been enough: marketing plans, attracting new customers, building market share all had the hallmarks of Andy's stellar success at Asda. But in the supermarket business you make the sale, the customer leaves and your relationship is usually over.

How different this is from selling a mortgage: making a long-term commitment in which your assessment of the individual and the asset is critical. Care is the key ingredient, and in the exercise of that care there should be no place for self-certification. Valuation should be strictly conservative. Apart from buy-to-let, which is a separate type of transaction and should be a separate class of market, a mortgage is a personal provision, not a business proposition.

In fairness, Andy's team's presentation that evening was on moving forward, removing the obstacles and tiresome procedures. Growth and speed were the watchwords; potential downsides were not on the agenda.

In studying the final collapse of HBOS, I could understand, if not accept, the surrender of value in the pursuit of growth in the mortgage market. But the reduction in the quality of the corporate lending book is not comprehensible; this had always been a strength. Did the degradation have its roots in the success of the Sears deal, I wondered, and did the confidence which that success instilled in the executive lead to its profligate lending? While I can, from my time in the bank, relate to the two classes of market, corporate lending and buy-to-let mortgages, I cannot comprehend the activities undertaken in America. What appears to have started as a treasury operation became involved in arbitrage using Grampian Funding. Grampian was an HBOS affiliate that had taken over a Jersey-registered fund named Pennine, acquiring $11 billion of assets in the process.

Grampian raised credit by issuing asset-backed commercial paper, borrowing short-term 90 to 270 days. With the money raised it bought mainly packages of American mortgages, paying an acceptable interest rate. So it borrowed short to buy long, reversing the fundamental banking principle that you borrow long and buy short. It came back to haunt them when the interbank lending market dried up and they could not meet their commercial papers' repayment schedules. Even worse, what they had bought were low-quality mortgages whose value was rapidly disappearing. These instruments, in fairness, did not seem to have been the toxic sub-prime variety, but they were almost as bad.

What I believe was initially intended to be a participation in the interbank lending market had become a class of market on its own, taking positions and pursuing profit where the bank had no history or experience. The transformation of interbank lending into a separate class was never presented to the board as a serious commercial opportunity in my time. I would like to know how it was reported to the board and how it was controlled; the results suggest a regrettable lack of governance.

To answer my own question, the seeds were planted in the mismanagement of the mortgage market and the departure of wisdom and judgment from corporate lending. Four key factors were, first, the tragedy of Alistair Grant's fatal illness. With him at the helm I am sure the fiasco of the American mortgage market would never have happened. Secondly, the departure of Peter Burt with his banking expertise and international experience. Thirdly, for various reasons, the corporate banking assessments were not as strong as they should have been, the banking roles were being taken by individuals who were unprepared for them and the regular independent assessments, such as that provided by Robin Browning, had gone. Lastly, Pattullo would not have approved of my involvement with Eddie George. He would have been the man to do it himself, and the fact that it had to be done at all was a red warning signal fatally ignored.

14

Canada: Sun Life and Canada Helicopter Corporation

Two separate involvements in Canada drew me into the adventures of two men, both large in physique, personality, mental ability and character. One was from Auchtermuchty in Scotland; the other from St John in Newfoundland. The opportunity to join the boards of their companies and others offered me an educational experience to engage with different problems in new situations. Each of them had attractions in addition to the international involvement I have always found stimulating, and contributing a view rather than a decision gave me a refreshing new perspective.

Sun Life

I met the first of the two diverse characters on a rugby field in Fife, Scotland. It was a tough, close match. I was a wing forward, so when the scrum was down I had to be ready to chase round and catch the scrum-half. I duly rushed round, head down ... and smacked hard into the chin of one John McNeil, knocking him out cold. There was no penalty, it was truly an accident; no foul play, it was just that my head was evidently harder than his chin. He went off shakily as I heaped apologies on him, and that was the last time I saw him for many years.

Apparently undamaged, he went on to become the chairman of Sun Life Financial of Canada, a titan in the insurance industry, and continued to build his business with the energy and intellect that got him there in the first place. My Shell colleague and ex-boss Peter Baxendell had served on his board, and when Peter retired he put my name forward to the nominations committee. After a number of interviews I was presented as his replacement. When John and I met again I looked at him carefully. He was bigger now but showed no scars from our earlier entanglement. A spell of unconsciousness had obviously not damaged his mental powers. So we shook hands and a happy relationship began.

In 1995 Sun Life had celebrated its first 125 years. During that time it had built a business in Canada that now employed more than 5,000 people, handling CA$3.7 billion of revenue and creating a net income of CA$164 million a year. In the US, revenues were $4.4 billion and net income $117 million in 1995. The UK had a smaller revenue but sufficient to earn $20 million net, but more interesting in a way, because it gives an insight into the corporation's vision and ambitions, was the $33 million income from the Far East: Manila, Hong Kong, Jakarta and Beijing. This sector had by far the highest growth rate at 42 per cent per year, and the market was more stable because the revenue was from life insurance and reinsurance.

In the Far East the company had not yet tapped into the other areas that were already developed in Western nations: annuities and savings, pensions, mutual funds, investment management and mortgages. With his international experience and feeling for countries that were just beginning to build their economies, John saw the opportunities. His Scottish upbringing had implanted an economic model that put children and education first, and home ownership and savings for that 'rainy day' close behind (it always rains in Auchtermuchty as the clouds sail over the Lomond hills). This was the opportunity which, much later, was exploited to great effect by Prudential chief Tidjane Thiam!

My experience of working in Thailand was relevant to these East Asian aspirations. I had been immersed in the region and its people and knew the challenges of growing a business there. In most countries where earnings were growing as economies developed, only a tiny proportion of the population was providing for their retirement, their health and their children's education. This offered enormous opportunities for this sector of the business in the twenty-first century—but it required courage and nerves of steel when it came to risk-taking. Happily, neither were wanting in John or his successor, Don Stewart, another Scot from the island of Arran. Don had been well prepared as the company's chief actuary, the man trained to know the risks and calculate the odds.

There was something in the mid-90s review that caught my eye. MFS Insurance Services, an affiliate in the US, had an annual growth of $50 million, 10 per cent more than forecast, and this had not been highlighted in the board report. Its mutual fund had increased by eight billion dollars to almost 50 billion dollars in the year. This American retirement market was growing and we were getting our share. I told John I wanted to go and see them. MFS's story seemed very like that of Capital Bank at the Bank of Scotland: it grew at high double-digit rates but the results were unrecognized at board level.

In Massachusetts we found a smooth, efficient, well-presented organization. The risk operation was being skilfully managed and the opportunities realised. The individuals were impressive and entrepreneurial, continually asking searching questions to identify vulnerabilities, and we agreed they should figure a bit more in our Toronto boardroom.

John had initially struggled to reconcile the derivatives market—options and swaps—with his Presbyterian upbringing, but in time he managed to suppress his puritanical inhibitions and accept the wisdom of using such instruments, which eventually put Sun Life in the vanguard of risk management. The company was now seen as the country's leading insurer. Its position was confirmed in December 1995 when it beat eight other strong contenders to win the contract for the government's public service healthcare plan, which covered more than one million people.

The ensuing visit from the regulator was—to me at least—a salutary experience. Early on the morning he was due, the whole board was assembled and the tension was palpable. John was subdued and Don distractedly sorted and resorted his papers. Nobody spoke, but there was a collective start when the door burst open at last. The regulator, a small man with a purposeful look and no papers, strode towards the assembly. He cut across the chairman's welcome, addressing no one in particular with, 'I have looked at the report and I don't like it. You boys will have to smarten up your act.' With that he turned on his heel and left.

Having taken a few moments to digest the oration, John spoke to a shocked room. 'Well, that was better than last year—and shorter! We'll get the report this afternoon and find out what's upsetting him.' Perhaps this miniature tyrant did not want a debate. Perhaps as he looked at the two Scotsmen present he knew that for every question there would be an answer, and for every criticism a solid rebuttal. Perhaps he had better things to do that morning than to listen to a roll-call of impressive achievements.

As an introduction to Canadian regulation it was brutal and rude. But for my colleagues it seemed normal and, as they rationalised, there were at least no threats. My education was under way. This was as far removed from the Thai silk-glove approach as one could possibly imagine!

The requisite smartening up was achieved by having a closer look at our international communications. Improving their efficiency made sure we were not missing opportunities in Europe just because of the time differences and distances involved. This satisfied the regulator—for the time being at least!

As John handed over to Don, my time with Sun Life in Toronto drew to an end, but my link with the company continued for another decade through the activities of Diligenta.

Canada Helicopter Corporation

Craig Dobbin from St John in Newfoundland was the second larger-than-life character in my Canadian story. As a boy, Craig had fished and cleaned fish for his father and learned how to work hard. He had had an outstanding career as a commercial property developer throughout the country.

Craig's life changed suddenly on the day that the helicopter he was travelling in crashed, killing some of his fellow passengers and seriously injuring others. Freeing himself from the wreckage, he set out to find help. He came across a railway line not far from the accident site and luckily, though the track was only lightly used, he was able to hail a passing train and summon help for his fellow passengers, thereby averting further disaster. Bears, which were in the area, are nosy creatures and always hungry!

As Craig recovered, he thought about the helicopter company he had been using and how much better it could be if he ran it himself. By the time he was up and about, Canadian Helicopter Company (CHC) was born, and it was soon a sizeable concern. It had one major customer: the oil business. Competition in price and service was strong, although the nature of the vehicles and of his main customer's geographical requirements gave the business a high catastrophic vulnerability. Flying in high winds and unreliable weather and landing safely on small platforms in wild seas was a real test of the pilot's competence. Craig took it a stage further and entered the area of air, sea and mountain rescue, which called for extended training and experience. Pilot expertise was at a premium and these skilled people knew their value, which made retaining them a challenge.

It was at about this point that Bill Stinson, a fellow board member at Sun Life, invited me to dinner with his great friend Craig. Bill had run the Canadian Pacific Railway, the railway that had tunnelled its way through the Rockies to reach the beautiful city of Vancouver. He and I realised as soon as we met that we had much in common, although from vastly different railways, mine having been on a tiny island with multiple intersections where we mixed passenger with freight—not to mention the huge commuting businesses.

The three of us sat down at the restaurant, and I listened. Craig, an irrepressible entrepreneur, told me how he had tackled the helicopter business in Nigeria and had built a strong relationship with a reliable local firm. When that partner wanted him to move into fixed-wing vehicles, Craig demurred: 'We knew how to sell service,' he said, 'but not tickets. Supplying the public is vastly different from dealing with one well-established financially strong oil company.'

The meeting was informative and stimulating, and I realised we had been talking non-stop for almost three hours. I had taken him through the helicopter accidents Shell had been involved in and given him an insight into the users' concerns, and he had listened thoughtfully, asking good questions. 'Well,' he said at last, 'the next board meeting is in St John in a month from now—why don't you come? Bring your wife; she will want to meet us.' So in that relaxed way I became a director.

Craig's company was growing fast and growth needs money, so he invited his bankers to his beautiful house in St John. Everything he did was spectacular, and the house was no exception. It was perched—and perched is the right word—on the edge of a cliff. You looked down from his garden onto shoals of fish and if you were lucky, a whale or two. Out in the open sea you could see icebergs in the near distance at the mouth of the inlet.

As one of the American bankers, glass in hand, was taking in all this natural grandeur, he called out, 'Craig, don't you have any ice in this house?'

'Yes,' came the reply, 'but you'll have to wait!' Within minutes we heard a helicopter take off and saw it head for the iceberg. It hovered over the glistening mountain, dropped a ladder from its belly and a man with an ice pick and bucket climbed down. In just a few more minutes, chunks of glacier were floating in our cocktails. Our drinks were well chilled that night—and Craig got his loans!

Five years later on a cold, rainy day in New York, Craig and I grabbed a cab, but no sooner had we jumped into it than he bustled us out, exclaiming, 'It's filled with smoke—I can't handle that!' I had detected a nicotine haze, which was annoying but hardly oppressive, so was surprised at his reaction. We continued on foot to our destination—a rather classy dive. In the days of Prohibition and the speakeasy, this place had had a hidden back room filled with bottles reserved for well-known political figures, actors and sports personalities—and the bottles were still there! In this bizarre setting Craig told me that he had only one original lung, the other being a transplant. He related how he had waited in hotels in mid-America with a personal jet on standby to take him immediately to wherever a new lung might become available. When the opportunity arose, the operation was done, which meant he could live, but it was essential he protected the new organ; hence the hurried exit from the smoky taxi.

We talked about the future and the challenges he faced. His great fishing friend, George Bush Sr., had advised him on a range of doctors, but Craig was philosophical. 'If the cancer comes back we'll fight it,' he said, 'but the chances of success are not high.'

Craig was building his business in Australia, Africa and Norway; taking his craft to the extremes of heat in the Niger Delta and the cold midday darkness of the far North Arctic; into the sandy fog off the Sahara, and the land of the midnight sun. But he was beginning to ail and as his business grew, his health waned. When he and his wife came to visit us in Woodstock, near Blenheim Palace in Oxfordshire, we spent a pleasant and sociable evening walking in the palace grounds. It was clear then that his health was fading, but his spirit was strong and his appetite for progress undimmed. Just a few months later when we were in Vancouver for the AGM, he was much worse. At dinner before the meeting I had persuaded him to go to bed as he was in intense pain. I suggested to his eldest son that he be prepared to explain that his father was unwell, and then take the meeting. But it was not necessary: Craig rallied, took the meeting in style, shook hands with all his shareholders and left in his private plane to return to his beloved St John home.

Within a few weeks I was in St John's historic basilica listening to modern popular music—'nothing serious' at Craig's request. We watched the final rites being administered and walked to the graveside accompanied by a ceremonial flight of CHC helicopters. When the earth was over the coffin, I returned to the house with Bill Stinson. Looking out at the icebergs standing sentinel on the inlet, we agreed how fortunate we had been to have known this man. He had lightened our lives.

Craig's boys wanted to sell the company. I knew from our conversation in Vancouver what Craig wanted for it and I held on obstinately until we got it. He had provided for everyone and I left St John as a memory, as mariners in previous centuries had left when the fishing season was over: a memory of human achievement in a challenging environment.

Although my time in St John was over, I stayed with the helicopter business in the UK. I felt I had something to offer CHC's successor, and they agreed. Assessment of and protection from catastrophic vulnerability stayed alive with me, and if I could help, I would. I owed it to Craig and to my own people who had given their lives in the waters of Sullum Voe.

15

International Petroleum Exchange: Digital Evolution

In 1999 the London Stock Exchange (LSE) announced plans to demutualize, ending its 240-year status as a member-owned organization and transforming itself into a company controlled by shareholders. In the same year the International Petroleum Exchange (IPE) took a proposal to an extraordinary general meeting to sell control of the organization to a consortium of five energy corporations. This proposal required a 75-per-cent-positive vote, which it did not achieve. As a result, the chief executive, Lynton Jones, resigned immediately, but the chairman, Peter Fraser (Lord Fraser of Carmyllie), agreed to stay on the board to guarantee stability. Having been a Shell executive in the early 1980s when the IPE was created, I had watched with interest from the sidelines.

Returning from a Bank of Scotland meeting in Edinburgh one evening, I was asked to divert to the Financial Services Authority (FSA)'s offices—a 15-floor glass mausoleum in Canary Wharf—to see Gay Huey Evans. I already knew Gay, an able American, straightforward and easy to talk to. I assumed the subject of the meeting would be the bank, since it was then becoming increasingly active and successful in the financial sector. In the meeting room all but one of the seats around the table were filled with staff, executives and advisers. I took the last place and Gay wasted no time coming to the point. She said they would like me to take over as chairman of the IPE. I paused, looking around the room at each person present, then said yes, I would do that, as long as they recognized that since I was now one of them, I could call on them for help if I needed it. They agreed, and we shook on it.

It was disappointing not to have had the opportunity to talk to Lynton Jones before he went. He had withdrawn immediately, which was unusual. I have never seen him since, nor had the chance to ask him why the demutualization issue was confused with a takeover bid: a two-step process would seem to have been a more logical course of action. In fairness to those members who were not involved in the bid, they probably were not keen on the concept of a consortium of competitors running their Exchange, and with my old Shell hat on, I would not have been either.

I had a high regard for Peter Fraser for his work in Scotland as solicitor general and his handling of the Lockerbie tragedy, and was delighted, as were my colleagues, that he agreed to stay on the board. The members of the Exchange had a deep affection for him—I noticed that even those he had censured gathered around him in animated conversation at the annual party.

The company's senior lawyer was another casualty in this imbroglio, albeit temporarily. Rather like Lynton, she seemed to take the loss of the demutualization vote personally, and was determined to leave, seeing 'no future' for the Exchange. She returned later, however, to play a constructive part in the business's growth and development.

I was charged with finding a replacement for Lynton, so the first move was relatively easy. Richard Ward was promoted to chief executive and David Peniket was confirmed as finance director. Objective number one was to revisit the demutualization vote. Looking around me at the first board meeting I noted that there was only one person who was not representing a corporation with a major financial interest: the independent member Mark Cutter. A successful amateur boxer, he was assertive, knowledgeable and articulate. In the years to come, he would be a reliable supporter of change and improvement. This conflict-of-interest situation would have to change, but that was a challenge for later.

Individual interviews suggested we had enough votes to pass the demutualization proposal but I needed certainty, so we identified the people who would carry the vote and followed up to ensure they attended and voted. As it transpired we had a full turnout and the decision was unanimous.

The next stage was to transform the Exchange from an enterprise owned by members into a corporation running a business for profit. Again the members voted unanimously in favour, satisfied that the proposal would ensure their participation in the exchange's subsequent conduct and direction.

When it was a mutually owned Exchange, 75 per cent of the members had had to vote on changes to the rules and on policy initiatives. In a rapidly changing competitive world this arrangement was too inflexible, and was anyway incompatible with a situation in which the board had full responsibility. With the new committee structure in place, the members supported our proposal to create a new board. Its remit would be to pursue opportunities in a market which, with the advent of the electronic revolution, was evolving fast.

Getting to grips with the business was not difficult. The information flow was well organized and the executives were clear in their assessment of strengths and weaknesses. The lack of capacity was a serious deficiency, however, and had to be addressed immediately. The open outcry trading floor would not be able to keep pace with the volume of trade and information flows demanded by an increasingly sophisticated clientele.

The increase in demand was due not just to economic growth, but also to the introduction and ballooning importance of risk management. The commercial world had already been badly affected by two oil crises, and understandably sought protection from further disruption. Proposals for construction work made assumptions based on fuel costs, as did airlines setting ticket prices for the holiday season and manufacturers setting prices for contractual deliveries to retailers. In the past their careful calculations had been voided by disruption in oil markets. By using the futures market offered by the Exchange they could insure against and therefore eliminate this risk. It was not a new idea: olive growers in Greece 500 years ago had sold their harvests 'forward', obtaining the funds to support their activities before the harvest took place.

*   *   *   *   *

We needed to find a partner who could introduce electronic systems that would integrate with those of our customers. The screen revolution was still in its infancy, but we were convinced it was the way forward. Were the integration successful, it would not only protect and grow the market, it would also be an important binding link to our customers, underpinning our market share.

We initiated the search and soon identified two preferred candidates. One was a well-established trading firm with an aggressive leadership, but we felt it did not have the new-generation approach we were looking for. The other was an upstart young entrepreneurial company based in Atlanta,

Georgia. Jeff Sprecher, chairman, founder and CEO of Intercontinental Exchange (ICE), was a professional in the business. His mastery of the detail and value criteria was outstanding. Moreover, as I sat listening to him in his sparsely furnished bachelor pad, facing a large undecorated concrete wall, I realised that here was another Lo van Wachem, living and loving his business. It was a decisive moment and I occasionally dream about the relationship created beside that wall.

Jeff's fellow director, highly competent and knowledgeable, was from Alabama. Like Jeff, Chuck Vice was fully conversant with the detail and a good people manager. He was a major contributor to the growing success of their business, and ultimately, ours. The third member of the group was the finance director, Richard Spencer. He was a fit, active soldier at heart, and a banker for Goldman Sachs. He did much to lubricate the deal and post-deal, successfully bringing the key players together. (Later, under President Trump, he became secretary of the navy.)

ICE foresaw the revolution that would overtake our activities. What they did not foresee, and neither did we, was the exponential growth in customers seeking risk-management cover through sophisticated hedging strategies and, later on, government decisions to move huge uninsured risks into safe havens.

Seeing how this company worked underlined how far behind the Petroleum Exchange had fallen, and made me realise that we had neither the technical competence nor the resources to catch up. So when the merchant bankers appeared alongside our preferred candidate with a proposition to merge, the way forward looked obvious. We had run out of steam and they were brim-full. 'Merger' was soon massaged into 'acquisition', but the deal included two places for us on the main board and worked along the lines of a major Shell affiliate company. I had no difficulty with the arrangement then, nor did I at any time during my subsequent 16-year chairmanship.

In 2005 we decided to become a public company through a share flotation. First we had to review the composition of the board of the IPE (which was shortly to be renamed as ICE Futures Europe) to make sure it complied with the corporate governance and accounting reforms and with the requirements set out by the FSA for a recognized investment exchange.

The IPE as originally named had an 11-person board, three of whom served on the ICE board. The IPE board's code of practice required consultation with members. Jeff and Richard represented the parent company's shareholders. The interests of member groups and other stakeholders were protected through independent committees—the authorization rules and conduct committee and the risk and audit committee—which also had direct relationships with the external auditors. The two key committees were set up, both with respected and experienced chairmen: Lord Fraser and Peter Nicholls, a long-term oil man.

The lengthy document covering these organizational issues confirmed ICE's understanding that the IPE needed to be able to operate as a separate entity with a measure of independence within the Group, though I never had recourse to this proviso. The document has stood the test of time for all stakeholders. Continuing excellent, if not outstanding, financial performance contributed to the good relationship and the respect that permeated the company at senior level was an important factor. Matters occasionally arose that challenged the status quo, but with our established structures and the judgment and balance of our board and committees, they were settled fairly and amicably.

On the trading floor, commercial transactions had historically been completed during periods of intense activity, which were followed by hours of inactivity. The floor members were a lively, articulate bunch who seemed to enjoy their work as much as their play. They helped each other through difficult times and generously supported charitable causes, but were clearly concerned about the future. They could see the technological revolution coming and knew their time was limited. Understandably it made them sad, though a number of them subsequently successfully guided their offspring into the world of screen-trading.

As the prospect of closure loomed, we set up training courses to familiarize traders with the technology that would take over the physical rough-and-tumble of the floor, but few converted. A personality-free screen with impassively blinking numbers was no substitute for the robust human contact and spells of hectic activity that had characterized their work. The death of the floor was not dramatic but, sadly, a natural expiry.

Before it closed, one unexpected event brought the traders together for a last hurrah. More than 30 Greenpeace activists in protest mode illegally invaded the floor with whistles, horns and panic alarms. The traders defended their territory with enthusiasm that rivalled any bar-room brawl. The police arrived to find their job had been done for them. Greenpeace did not apologise for wrecking the floor, but the activists never came back.

*   *   *   *   *

We needed to integrate more closely with our customers in order to introduce screen trading. There were agencies that could facilitate this, but I was not comfortable with that idea. We did not want a series of one-off implementations, a patchwork that would have to be revisited continuously—we needed a common set of procedures. Setting up our own processes was not simple initially, but it made future development infinitely easier. We knew our systems because we had built them. First hand, not second hand, is always my preference.

Customer-driven product opportunities increased the market volume, and before long logjams built up. But they were not in the contract design and approvals area, they were in the clearing house, where there were delays in accepting new products. This was holding us back. As the queue of products awaiting approval grew, so did our determination to find an alternative to the London Clearing House (LCH). Its board was not effective in dealing with operational problems, and its members were powerful and difficult to work with. Any merger would require strong executive action from the clearing house, which a conflicted board would find hard to support. Many years later these factors led me to advise the ICE chairman to steer clear of involvement with the LSE Group, which was only a part-owner of LCH.

At that time there was no alternative clearing arrangement, so we would have to create our own. We appointed an executive with the energy, knowledge and creativity to put a new house in place: Paul Swann, who had served an apprenticeship with LCH. He assembled his team and they set to work putting control structures and supporting information flows in place. LCH was cooperative and we agreed a handover date.

The weekend before we were due to transfer margins, I was on a hill in Oxfordshire showing my grandchildren the ancient chalk horse carved into the grassy slope when my phone rang. It was the FSA. The executive in charge of the trading and clearing section said she could not give details, but Lehman Brothers was in deep trouble and might well go down. If it did, the repercussions would be extremely serious and could significantly disrupt our activities. I said I would put an immediate hold on our transfer, and with another call instructed the executive to suspend it on a technicality and reset the date for a couple of weeks hence. No reference to Lehman's was to be made; in the wrong hands that information could have had disastrous financial implications.

Three weeks later the margins were transferred with minimal disruption; the backlog of approvals began to disappear immediately, and the business grew. The volumes and rivers of money surged like the Niger watering the mangrove swamps of Eastern Nigeria: year-on-year growth soared as the commercial world, in the aftermath of Lehman, became focused on risk mitigation.

Banks liked this shift in emphasis and became expert in it, investing for direct market gain. Proprietary trading, as it was called, became a bête noire for Paul Volcker, the Federal Reserve chairman who is credited with ending high inflation in the US during the 1970s and early '80s. He believed it was an area of prime speculation that banks should not be involved in, and his view finally prevailed. Mr Trump may have had second thoughts, but Volcker's position was that the main arteries of the financial system should not be compromised by daily transactional risk-taking in the name of profit.

Many, if not most, of the financial deals across the transactions board were completed 'over the counter' (OTC), that is between two parties with no third-party underwriting. Their volume was growing at an accelerating rate as the world of commerce expanded in all directions. There was no longer a single-line business from loading port to destination port; there was a complex network of interlinked activities, physical and financial, conducted throughout the voyage. Despite this, the financial arrangements were carried out on an almost informal basis. There was no insurance or margining system to ensure completion: this was systemic risk in spades!

After Lehman and the sub-prime assets scandal, the regulators were looking for the next risk to the financial system. Since OTC business had no insurance protection against catastrophic events, they thought that would be a good place to start. With a stroke of the regulatory pen they changed the system so that it now had to be cleared and have a margin in place to insure it. As a consequence, clearing houses inherited—or rather were endowed with—all OTC business. They instantly grew significantly bigger, became a key part of the defence system and a focal point for regulatory activity. The river, which had already been in spate, surged.

*   *   *   *   *

The root of the Lehman default lies back in history before its fatal crisis. The first sign of difficulty in the US financial world came in March 2008. I recall it vividly, as I was involved in a bizarre game of golf in Florida after an ICE board meeting. Two colleagues invited me to play at their club in Naples at a course where Luke Donald (PGA Player of the Year 2011) practised for his American tournaments. Jo was visiting friends in America at the time, and had also been invited.

Golf in America is not a walking game: golf carts are the order of the day. In my view this defeats the purpose, which is healthy exercise for all parts of the body but particularly the legs. Walking also gives you time to contemplate the poverty or excellence of your last shot, and the chance to plan how to do better with the next one—rather like business!

Our friends seemed distracted, and uncharacteristically wanted fast action. Golf has a natural resistance to speed: the golf swing does not do rushes. But here balls flew about in all directions—into bushes, bunkers, water—until at last one reached the green. We were making record time, if not record scores.

At about hole six we came up against a relaxed four-ball. We apologised and shot past them, then on to the eighth, having missed the seventh. With three tee shots despatched we were back on fast track, my colleagues' phones ringing insistently all the while. One hole had a remarkable feature. The tee was set alongside a fence, beyond which was a two-storey house. From the upstairs window a chute descended steeply into a pool. 'That's living!' remarked one of my hosts as he hurried on. We sped past all and sundry until we arrived at the 18th. As we finished and shook hands, my partner apologised, saying they had a 'serious problem'. As they then hastened off to the car, they asked if Jo and I would please wait for a man who was coming to see them, tell him they had gone on ahead, and ask him to hang on until they called him in an hour's time. So we made our way to the clubhouse, returned the rented clubs and waited for our car.

A limousine soon drew up and a largish man carrying a coat was ushered out. I went to meet him, but with a great sense of urgency he strode past me and up the stairs, handing me his coat as he went. This was no ordinary coat—nothing was ordinary that day!—it had been immaculately tailored from the finest cashmere. He was surprised to be introduced to Jo, who he assumed was some non-executive director he was unaware of.

When he realised I was not a club employee but was there to give him a message, he calmed down. He explained he was a consultant and had been invited to restructure Bear Stearns. Suddenly all became clear. My golf-playing colleagues were directors of the company and had told me that it was on the edge of collapse. We bought the man a drink and he told us he lived in hotels and drove only rented cars. He had no tangible assets (except the coat) and no living expenses, moving from job to job on somebody else's ticket. We left him when our car arrived, trusting our colleagues would pick him up later. When they did, believe it or not he forgot his coat, leaving him even less asset-heavy!

The episode at the golf club was comic; my colleagues' distress of was not. Bear Stearns was enmeshed in the fabric of the United States' financial system, with 400 subsidiaries and 5,000 counter-parties throughout the world. It had borrowed $80 billion in the repo (short-term) market, and one-third of the collateral was in mortgage securities. If Bear Stearns should fail, the repo lenders would unload, depressing the price of these securities and that of everybody else's collateral.

This kind of systemic risk spreads like a virus. Too many firms depended on short-term funding and exposed themselves to huge losses in the housing market. For Bear Stearns, it was too late for restructuring and too early for help from the financial authorities, who were only just beginning to realise the scale of the problems. After government pressure, JP Morgan agreed to take the company over on terms that were workable—but disastrous for Bear Stearns' shareholders.

*   *   *   *   *

Home ownership is the common starting point for a family's investment, and America had the technical capability to provide a growing stock of property. The assets would be financed by mortgages extended to the buyer, and contracts guaranteeing repayment of the principal and regular interest payments would be returned to the lenders as collateral.

As the housing sector grew the prices rose, along with the financial community's ingenuity. Skilful financial engineers packaged the mortgages and in some cases split them to create attractive bonds to be floated on the market. These bonds were highly rated by compliant rating agencies, thus giving them a badge of respectability. No one seems to have thought about the vulnerability of the basic premise: the loan was given on day one, the interest on the loan was to be paid over the long term at regular intervals with the mortgage repayments. It was a long-term loan supported by short-term financing—exactly the opposite of the fundamental banking rule that ensures a transaction's viability and liquidity: borrow long, lend short.

The whole concept depended on the individual's willingness to pay. This I had real experience of, starting at the age of nine when customers could not pay their butcher's bill after an expensive Hogmanay, and again 25 years later when an army would not pay its fuel bills because the country was at war. Already I had noticed rumours of a housing bubble in the US press. This was ominous. If prices collapsed, liquidity would drain out of the system. Meanwhile, the customers' new purchases were quickly losing value. The market had no interest in rescuing something that would be cheaper tomorrow, nor of course had the homeowners, who handed back the keys along with the mortgages! Bear Stearns was only the beginning.

The contagion spread. In 2005 and '06, Lehman Brothers had issued $106 billion of sub-prime mortgages, Royal Bank of Scotland $99.3 billion, Countrywide Financial $74.5 billion, Morgan Stanley $74.3 billion and Credit Suisse $77.4 billion ... and they were not alone. These amazing figures illustrate the extent to which the market's appetite for high-yielding investments suppressed thoughts of risk—or indeed reality. Incredibly, traders showed no understanding of, or even curiosity about, the health of the underlying assets. This wild behaviour swept on, well out of control, until eventually it took with it large and serious corporations. It became the subject of books and films in which reality far outdid fantasy.

A stampede away from reality is not unusual. There was one in Lagos in the 1970s that produced a price of $42 per barrel in a market where $12–15 was the norm (though nobody lifted any of it). In 2007 Northern Rock was offering 95 per cent mortgages with an additional 30 per cent on top for home improvements. Their decision-makers had lost touch with reality, concentrating on marketing while ignoring financing fundamentals. In that case, it ended in tragedy and a national disaster.

The politicians' response to the crash and the injection of billions of dollars into the financial system was to produce more regulation and prescribe the establishment of more procedures. This looks effective, but includes nothing that will check or monitor the forces that could create the next crisis. Opportunities arise in periods of change and disruption, and are a gateway to wealth creation—honest and dishonest. You need look no further than Brexit for the next opportunities.

My response to these calamities as an executive is that, particularly in the financial world, we need to prepare key decision-makers to ensure they are exposed to the whole of a business proposition, not just a single facet. In the Bear Stearns case, the marketeers and financial engineers dominated, slicing and dicing mortgages with no reference whatever to the financial implications. The oil industry had the same myopia in some companies: market share prevailed above all else, and price was the mechanism to achieve it. Marketeers knew their target, and the rest was for the finance department to worry about. Personal development within in a company should continually train decision-makers to see the whole picture; to understand that the health of the business is the prime objective.

After Bear Stearns disappeared into JP Morgan, the focus was on Lehman Brothers. The United States' federal authorities looked for ways to save a system that was teetering on insolvency. They tried to repeat the Bear Stearns disappearing act, this time with Barclays in the UK attempting a Lehman takeover. But Sir Callum McCarthy, chairman of the FSA, and Chancellor Alistair Darling had a fragile situation of their own to deal with and made it clear that such a purchase by Barclays would not get through the government's regulatory approval system. Bank of America also declared it would not take $70 billion worth of toxic assets from Lehman, so interest faded there too. The only option left was for Lehman to file for bankruptcy, which it did. The federal authorities, seeing no other way out, let it happen.

Lehman's bankruptcy did not solve the United States' problems, although the Bank of America's subsequent takeover of Merrill Lynch for $50 billion did something to steady the ship.

American International Group, known as AIG, the anchor insurance company with a trillion-dollar balance sheet and 115,000 employees, had a subsidiary called AIG Financial Products, which had sold insurance against the risk of a housing slump. As the situation in the housing market deteriorated, AIG was besieged with margin calls on the entire range of its mortgage security products. The rating agencies were aware of this and were threatening to reduce AIG's credit rating, which would force the company to post billions of dollars of additional collateral. AIG was now added to the list of casualties following Lehman's bankruptcy, making another problem for the financial authorities to solve. And solve it they did, with government financial aid, negotiated through a tough agreement. While this support may have offended some purists, it worked, and in those frantic days this was enough. Stones were gradually being wedged back into the dyke.

As an economist, I watched with fascination. The crisis produced government intervention in private sector affairs in a free-enterprise economy that would have been inconceivable in earlier years. Also the encouragement of 'attractive' acquisitions—as with Bear Stearns and Merrill Lynch—was a new tool in the supervisors' locker. The scale of the crisis was frightening. It was rooted in individual citizens buying homes with technical mechanisms that, because of their disconnection from the real market economy, supported an absurdly unreal volume of activity.

*   *   *   *   *

In contemplating this frenetic chapter in economic history, one can make certain observations, the most fundamental being that sound decision-making is dependent on seeing the whole picture. In a world where objective-setting is narrowly defined and well rewarded, attention can become too focused on isolated targets. But even this cannot explain the madness of sub-prime. Was the citizen investor's appetite for products that would earn far more than US Treasury Bonds insatiable? Had the creation of the asset-backed mortgage security, priced by the official rating agency, been so attractive that it lured buyers into looking no further? The answers must be yes. In this fantasy period the technicians took over, dividing assets created for the rating agencies.

Waking up to reality is painful and takes time. I first discovered this in the 1974 oil crisis. Until then, remuneration for discovering and producing oil went to the corporations that managed the fields, and they in turn paid their taxes to the host governments. OPEC (Oil Petroleum Exporting Countries) then emerged to change this, the oil price subsequently becoming subject to agreement with the national authority. OPEC's regular meetings and commitment to controlling levels of production heavily influence prices. But that influence is uncertain, because each producer has its own political and economic concerns to balance. Rhetoric is seldom matched by action, so the market continues to be difficult to read.

For me, this reality left its mark. The wisdom emblazoned on the front of Nigerian mammy wagons, 'No condition is permanent' is right. In oil, as in finance, you had to recognize and understand the forces at work in order to survive.

While the United States went through this traumatic period, we concentrated on keeping our operations solid at home. We were ready to dismantle and distribute the Lehman portfolio among members, and did so speedily after bankruptcy was declared. Thanks to careful planning and preparation, experienced manager François Lepart completed the whole operation in two days.

Our clearing house was a step change in UK activity. It was no longer just the completion of the contract that concerned us, it was also a validation of the participants' financial viability. A clearing house clears, settles and guarantees the financial performance of futures contracts and options on futures contracts, and clears credit default swap instruments.

Crisis-management tests are carried out regularly with the members, in which the defaulter's (theoretical) portfolio is auctioned and the root problem identified. This may seem theatrical, but the procedure reminds everybody that risk is real, and that there is a process in place.

Individual members' fortunes change all the time, reflecting their portfolios' contents, the risks they have taken, and their successes and failures. The clearing house must maintain a continuous review of its financial strength. It does so by loading relevant data onto a risk matrix that quantifies the severity of any financial concern and calculates its potential impact on the system. This is a key management tool which is under constant review.

In my years on the board of ICE, my daily preoccupation was managing and directing the exchange as a state-of-the art financial machine. It provided a fast and secure transaction facility for its customers and a clear picture of the market and its potential pitfalls for the directors. It examined the members' viability well enough to have put Lehman on the watch list well in advance, and alert JP Morgan to the activities of its rogue trader, the 'London Whale'.

Each of the markets we were involved in was subject to forces specific to their environment. For oil it was politics; for cocoa it was weather; for gold it was the jewellery market and financial security; for stocks and shares it was the mood and work of a disparate range of investors. Identifying the particular force and then monitoring it was key to protective action.

Our corporate objective was to use our financial strength to expand and underpin our business. Acquiring the Singapore Exchange, a long-established trading point used by China, was a natural and sensible out-step. It took us into the areas of the Asian powerhouse without entangling us with an unfamiliar environment. I had long experience with Singapore as a refining centre and as a market, and it fitted well with ICE's trading model. Acquiring the New York Stock Exchange was another attractive and well-priced investment as well as a valuable learning opportunity.

History had warned us how politically sensitive and vulnerable to policy changes our markets were. We needed to analyse our position and look outside it to see the full picture. Having moved all our business on-screen and perfected our processes, we had created a second opportunity in the instant availability of transactional data. Investors now had a fund of live information on which to strategize and act, and regulators had real-time data to monitor. The design and management of this information flow was a new business, and one that we became more actively involved in when we acquired Data Corporation in 2015.

Data collection and distribution has its own impact on the volume of activity. It also informs the decision-makers and alerts them to the risks involved. The outcome of their informed decisions will, in the 'new' data world, be accurately recorded, but overall success remains a matter of personal judgment. For those of us who entered the trading business in the early days, this is the true excitement.

16

Business Opportunity: Diligenta and Jubilant

I first came across Tata Consulting Services (TCS) when its imaginative approach to digitizing accounts solved a pressing problem at London Electricity in double-quick time. Some years later Executive Director Natarajan Chandrasekaran, known as Chandra, updated me about TCS's activities in the UK and asked me to join the board of its affiliate company, Diligenta. I agreed, first because I liked and admired the TCS people I had already dealt with, secondly because I believed that the organization was, like Intercontinental Exchange, riding a digital wave which in time would revolutionize, modernize and streamline the structures underpinning their area of the financial industry.

Diligenta's target was the insurance industry, where millions of personal pensions were in the making. It had established that the data-management systems used by the major companies were all manual, if not manuscript. The first stage of its plan was to computerize this mass of data and make it accessible to the executive in a common format in real time. The first big contract was with the Phoenix Group. Contract negotiation was tough and I thought the service-level assessments, which clarified what was expected of the provider, were set unrealistically high. The state of the data also presented a real challenge, but Diligenta rose to it.

The staff who came with the data as part of the deal also faced new challenges: the introduction of new people, new procedures and new processes into an established bureaucracy. Implementing change requires patience and persistence, but keeping the team leaderships together meant that the lessons learned were carried forward into future contracts, each success resulting in a stronger team. It was a learning experience on both sides, and on balance both did well.

Streamlining communication by using electronic messaging was not by then a revolutionary idea, but it did require the customer to be prepared and to cooperate, and the operatives to be trained. The timing of the change was helpful, since by then many policyholders were already computer literate. The skills acquired in that first contract stood the team in good stead for future contracts, including the negotiators, who subsequently agreed more achievable service-level aspirations.

The Diligenta team began by systematically analysing complaints and incidents and setting targets for improvement, which they presented to the audit committee and the board. This helped define the best way to satisfy customers. Establishing customer relationships became simpler, quicker and significantly more cost-effective. Demands on customer time reduced significantly, and this continued as the change process matured.

The improvements affected not only the policyholders but also the corporate clients, who began to see opportunities in working more closely with their insurer. We, the providers, could customise contracts with realistic and real-time terms, so were writing ourselves into the fabric of the relationship. We had achieved something very similar at ICE when we entered the digital age. It started with a contractual relationship but melded into a partnership where products were created to meet specific needs, and where monitoring was shared and understood.

Financial systems, however, be they in energy, commodities or financial products, are subject to manipulation and misdemeanour. Cyber-attacks were and are a reality, and the perpetrators' expertise and ingenuity can be astonishing. The clearing house or contractor plays an essential role in defending against such abuse.

The initial contracts with insurance companies were set for lengthy periods, ten years with a break clause if required, but in reality they were evergreen contracts. In this 'closed book' business, benefits become available when contracts naturally terminate. A change in regulation in 2015, however, permitted customers to access benefits earlier than originally planned. Take-up of this change has been slow, reflecting beneficiaries' caution—and perhaps the absence of any significantly attractive option. For the provider there is a dilemma: should they encourage customers to hold on up to the normal termination date, or offer new investment opportunities to promote early redemption? The latter is potentially hazardous, since opening your customer up to competitive activity risks losing a long-term profitable relationship. Policyholders might also feel their financial future is better safeguarded with the original policy, which (one hopes!) would suggest they have some confidence in the insurance companies' management of their accruing wealth. For Diligenta and others, declining books present a challenge, and the pursuit of new business will become critical in the future.

Jubilant

Jubilant is an entrepreneurial group founded by the Bhartia family. Their investment in pharmaceuticals and life sciences, with a sales turnover of more than $350 million, was the founder group. They capitalised on their retailing expertise through the Domino's Pizza franchise among much else, but were also tantalised by the opportunities of oil and gas.

They had no difficulty working in partnership with experienced investors, both private and public. Seeking operating experience, they asked me to join their exploration and production company in 2007. The Bhartias were not passive investors, they liked action. So very happily I found myself on a plane heading for their only oilfield in Kharsang, Arunachal Pradesh, India. Flying along the Himalayas as the early morning sun lit up the snow-capped peaks was a heavenly, unforgettable experience. All of northern India lay beneath us, stretching from the broad river at the foot of the range to the glinting, dazzling peaks.

We drove through the tea gardens until a towering escarpment loomed above us. Within the escarpment, I could see thick seams of coal being picked at by large mechanical diggers. Then on through the forest until we reached a small town, beyond which Jubilant's concession started. The drilling site was well ordered and all the pipes laid out. We had dug deep, but so far it looked like a dry hole.

As we toured the area, we noticed slicks of oil that had leaked out of the underground reservoir and found their way to the surface; with the coal mountains towering above and around us, this certainly seemed to signal hydrocarbon territory. But our discussions with the site engineer about that well and the earlier drilling efforts were less encouraging. The plan was to go deeper into the current hole, but there was little optimism that that would yield results. We discussed the coal and oil slicks around us to try to understand their significance. It seemed to me that the hydrocarbon had, over millennia, been compressed into leakages, but we needed the view of a geologist, someone who could compare our conditions with those in other areas that had similar characteristics. It made little sense to go deeper without further investigation.

Cairn, an independent oil and gas exploration and development company, had an offshore site that seemed to have potential, and this plus other sites in eastern India where it had found oil made a promising-looking portfolio. Getting involved with them would have been an attractive proposition had the price been right, and the oil man in me ached to forge ahead, to find a way to seize the opportunity. But my friends had not consulted me to share the excitement of exploring impenetrable forests, taming hostile territories and bringing home the prize whatever the cost—they were looking for an objective analysis. There were real, practical obstacles to consider.

As we worked our way through the options, I visited a potential concession close to the Burmese border. It was near the battleground where Field Marshal William Slim fought his decisive battle against the Japanese, slowly dragging them further and further away from their supply bases. The area was close to dense jungle and I helicoptered over it for four hours—which did little to suppress the old instincts! The roads and flimsy bridges would make communication the first obstacle to prospectors, and the challenge of firing charges would be the next. What the local police and military would think of seismic teams using explosives in a politically sensitive area was another question. All the environmental conditions suggested there would be a reservoir, but stacked against this possibility were too many unknowns. The financial conditions were simply not attractive enough for the potential investors—either that, or they just had too much common sense! Just as in British-Borneo, we were out of our depth. It went sorely against the grain, but ultimately I had to agree with the shareholders that there were better investment opportunities for them than oil-prospecting in deepest India.

17

Reflections

On the evening of my final day of full-time work I reflected on how I had spent the last 60 years. The three major occupations had been with Shell, British Rail and Intercontinental Exchange: vastly different entities but with elements in common. The oil industry and the railways are both hazardous work environments, Piper Alpha and Clapham Junction vividly illustrating the appalling price they can exact.

Some recollections from those years are so clear they could have happened yesterday. The battle with escaping gas on the beach in Borneo; the gas pouring from the well in Nigeria that drifted towards burning flares while we watched, helpless … until instead of exploding and blowing us to smithereens, the ice-cold cloud snuffed out the flames as one would snuff out a candle, and we breathed again.

It seems to me that Death has no preference. It can come on a cold wet night to an engineer working on the track, an arrow of steel speeding through the darkness, wind and driving rain muffling its approach. It can come from a helicopter falling into the chill and turbulent North Sea, and carry bodies 100 miles away on the confluence of currents. Church services I attended near the dangerous waters west of the Shetland Islands paid tribute to those who had lost their lives, and prayed for the security and safety of those who continued. This was the recognition of reality. But we human beings have the intellect, resilience and creativity to arm ourselves against these elements.

In the physical and mental challenges individuals face on the line, platform or plant, money and profit are not the drivers. Reward has its place but that comes later and is off-site. People's dedication to continued achievement is what makes working with them so motivating and attractive. For those in the railways, each journey safely completed is a victory, and for the geologist, the geophysicist, the driller and the engineer, each finished shift is another win against the elements.

The dangers lie not only in the activity but also in the process. Speeding without care, seizing opportunity heedless of consequence, ignoring irksome procedure to chase deadlines, these are always temptations. The conduct of the activity, whether it be locomotives, computers or high-speed drill bits thousands of feet below the seabed, depends on people. Investment in those people through careful preparation and management is the best way to ensure success.

*   *   *   *   *

With success there can be further temptation: the opportunity to take what is not yours. As Sir Alec Ferguson told the students of Robert Gordon's University in Aberdeen, the opportunity for transgression (a Presbyterian euphemism implying movement when it should define The Fall!) comes all too soon and too often. My experience in the butcher's shop taught me the human tragedy of temptation. A colleague stole a half-pound of mince: a minuscule crime and gain compared with illegally fixing trades or wrongly awarding contracts, but my disappointment in that friend at that early stage in my life was no less bitter. It was like losing a hand—what had been done could not be undone. He was sent on his way, out of my father's shop and out of my life.

So it was with the two traders who went to college together, joined separate firms and combined their skills to rig transactions to their advantage. Over two million dollars later, two promising careers were blighted. But it is not just the young and inexperienced who fall: witness the top manager who, after a hard-negotiated deal in which he had played no part, decided he should take a cut. It was only his sudden display of affluence that gave him away, and on that occasion we watched with satisfaction as he disappeared.

The success of our flexible, fluid trading system at the exchange and clearing house allowed financial firms to manage their risks while participating in market movements. Correctly read, these movements could generate huge profits, which made the operators extremely valuable. Their expertise was tradeable, so whole teams would change employers, generating staggering levels of remuneration and often, unfortunately, staggering levels of arrogance. They had become an active ingredient in the lucrative process without having invested a penny in it. The increased profits affected the profits of the bank that had financed the corporations involved, which then as a consequence increased the top banking executives' pay. It did not go unnoticed that banks were dangerously exposing their financial integrity to commercial transactions, taking risks in what were exceedingly volatile markets.

* * * * *

When I started at the railways, I was surprised at the high number of drivers who had chalked up over 40 years' service. Their experience was a real strength. They were the salt of the earth, rising early in the dark to take out their trains for morning peaks and mainline services, hauling people to work and coal to power stations. In the oil industry I ran 24-hour operations staffed by engineers, craftsmen and drillers on steel islands probing the earth's substructures, and refinery operators distilling products for motorcar and lorry. In the exchange and clearing house, my staff ran systems that underpinned the country's financial infrastructure and its international affiliations. These colleagues were, with only very few exceptions, first-class people.

Within these groups, there was a harmony brought by achievement and maintained by management. It was my job to ensure that this continued, despite the vicissitudes of the elements, the restlessness and the periods of frustration brought on by regulatory restraint. This required active presence and total availability.

In large organizations it is vital that employees know that you are accessible, not locked up in some remote office protected by gatekeepers. Communication is what keeps leaders alive to staff concerns; interpreting the sometimes-chaotic information flow allows you to feel the pulse and—hopefully—produce the right decisions.

Public appreciation is also important to those who work in these industries, even though those on the outside cannot have a full picture of what goes on inside, no matter how much information is published. This is why morale is vital. Morale can be nurtured by communication and a proper understanding within the company at least of what's being attempted and what's being achieved, and celebrating success is as important as sharing grief when colleagues are lost in hazardous operations.

Consent in an organization is also essential. By consent I mean commitment to engage. It is more than acceptance; what you are looking for in managing large and small groups of people is their buy-in, their willingness to hear and act on the message. To assess whether or not you have consent, you must consult widely, and you need the—sometimes painful—experience to know what questions to ask. An informal follow-up discussion I had long after an amalgamation between two strong organizations provides a good example. In talking to individuals from both firms ten years after the fact, I realised that they still thought within the parameters of their previous incarnations. There had been no real coming together, so the asset that would have been their combined strength had never been realised. That lesson needs to be absorbed into personnel and individual development planning if companies are to reap full advantage of carefully laid plans.

*   *   *   *   *

If you are the chairman and chief executive, the person you have to assess constantly is you, and you do this by comparing yourself with individuals in similar situations. You need to identify the ingredients of success as well as the decisions and behavioural lapses that lead to failure.

At the head of an organization you have a structure reporting to you which in turn has its own heads and hierarchies. Your relationship with these heads is key, and it may not be easy, particularly if one of them believes that your position should have been his or hers. There can be a temptation to reduce face-to-face contact with such an individual until your relationship takes on a simple question-and-answer format. From there it can degenerate into a written exchange, removing personal contact entirely. Email, unfortunately, can be complicit in this corrosive process. If you reach that point, every policy or development becomes a contested area, examined from a 'closed mind' stance, and discussion and objective review are impossible. It is critical to resolve such situations, even if it means reshuffling—or removing or replacing—the individual to fix the dynamic. Such an extreme solution must be a last resort, though, as it doesn't go unnoticed within the group and it touches your reputation too, which in turn affects the other relationships you are building.

There is a danger of isolation in a leadership role, so the review process must be carefully designed to guard against it. The wider your circle of consultation, the closer you are to reality and the surer your responses can be. Consultation can often be easier in social, informal situations when people are less inhibited in their exchanges. You need to be able to reach deep into the organization and feel where the resistance to or acceptance of change is strongest, and in that way to identify opportunities for progress. If there is an inherited hierarchy, you must work within it until it ceases to function positively, at which point you must contemplate radical change—and then act swiftly! Organizations can read your intentions more quickly than you might think, and this is where your being accessible is crucial. Your message must be clear, and it must resonate.

It can be easy in these rough waters to seek consolation from those who work closely with you, and this can develop into associations that have repercussions in your personal life. Allowing that to happen is the essence of stupidity. From one problem you create two, the second bringing wider and far more hurtful consequences than the first. No amount of rationalisation can cover the fact that such adventures emanate from a failure to cope, and this is what should be addressed at the very first signs of difficulty.

The fascination of leadership is the opportunity it provides for working with people. It is the release of their talent and their energy that brings success, and your capacity to orchestrate this is the true measure of your achievement.

So to the inevitable question: Would I do it again? The simple answer is: Yes. My father, a soldier at 16 in the trenches in France, in the fields of Turkey and in the final battles in Italy, never failed to support me with common-sense guidance, and he would not have accepted a 'No' to that question. Later I worked with two other soldiers; one who, with his Australian colleagues, won back Brunei for the Allies, and another who was caught in Dunkirk. Both were impervious to rank and both instilled in me the importance of assuming there was more to any situation than met the eye.

Borneo, Nigeria, East Africa, Thailand, Iran, Australia ... none of these would Jo and I have missed. Our expatriate years were full of the excitement of the new and the challenge of building yet another home together, creating a place where new friends and colleagues would come. The background—and at times a foreground—of revolution and civil war that coloured our lives' journey set a scene not so different from that of the World War that coloured our childhoods. But now we were much wiser and more closely involved—if not in the action, in the aftermath.

Without relatives around us we had an independence that forced us to engage with these new environments, to participate in our surroundings, and this was true even in our final return to London. It was a challenge we both welcomed. We realised it was truly a political business we were in, so we set out to understand the history of whatever country we found ourselves in, its political evolution, and the strength and weakness of its economy.

Nigeria first tested these fine resolutions, but for me it was not easy to fulfil them, at least initially, because of the all-absorbing work ethic of an exploration production company. Jo had more opportunity to engage widely through her educational work and her societal involvements, and her connections, her conversations, her pronouncements brought the real Nigeria into our living room. And so it was wherever we went.

It was also important that our circle of friends and acquaintances in that new country should be wide, and that we should listen to and understand them—even if we did not always agree with them. For this reason among many others, Jo's academic bent and her ability to sift facts from biased presentations and reports was invaluable.

After years of intense physical activity abroad came the world of money, and here the melding of a moribund exchange into a dynamic corporation was a recognizable success. There were of course financial rewards, but still it lacked the excitement of the offshore platform or the opening of a highway under the sea with the Queen to bless it.

Old dogs like old things.

*   *   *   *   *

As we reached our 80s and passed on many of our direct responsibilities, we had to acknowledge that our front-line time was over. The intellectual challenge was still there for Jo with her studies and charitable involvements, and for me there was still some commercial interest and other activity, but there was a growing dissatisfaction with the lack of progress in our political environment. I have wondered whether, one night over 40 years ago, I made the wrong decision not to be involved in my country's politics. I had been offered the Liberal candidacy of East Fife, which had been a Liberal stronghold since my grandfather's time. It was a beautiful summer's evening and the moonlight was shining on the Firth of Forth—it felt like a siren call to come home. I thought long and hard about it, but after all I was happy in what I was doing. It had its risks, high risks at times, but that added to the excitement. Jo was content with her life and would support whatever we decided to do as long as we did it together. The die was cast. Was it the right decision? Yes. For 65 years I shared an appetite for new adventure in distant places with a special partner who, like me, had no regrets.

Joan Reid

Joan Mary Oram was born on 18 October 1932 in Calcutta. Her mother, Dr Katherine Blackstock, was married to John Oram, a businessman from Dundee. Both went to St Andrews University. Katherine's two stepbrothers, who had run the Irrawaddy Channel Service and built a successful business in Rangoon, supported her studies. Pioneering was in the blood and these brothers, both maritime engineers, would have applauded the journeying of their step-niece, Jo.

By the time Jo was five years old, the threat of war was growing. The family returned home to Manchester, but within three years the threat had become a reality. Katherine was constantly on call and John, in charge of a large food manufacturing firm, spent practically every night trying to safeguard the roofs of his warehouses as bombs rained down on the city.

It was an exhausting time and eventually the couple decided to find a more suitable environment for their family. They sent Jo, the eldest, to Cheltenham College, soon to be followed by her two sisters. Her brother was despatched to Uppingham. All four siblings of this highly academic family followed their parents to St Andrews. Jean became a research scientist and Sally a nursing sister, while their brother, Bill, became a mathematician.

St Andrews was an inspired choice for Jo. Among very bright students and top-tier sportsmen and -women, she excelled all round. She became a first-class lacrosse player and in time won international honours. A talented tennis player, she qualified for Junior Wimbledon. She was, however, at pains to downplay her sporting achievements, even when she won the Ladies' Golf Championship at the Ikoyi Club in Lagos, Nigeria. But she could not deny the six fours she hit as a 'mother' playing against the Dragon School First Eleven, at which stage the umpire dismissed her as being in the wrong place. He argued she should have been up the road playing for the county team!

At St Andrews Jo progressed with little difficulty, her literary skill and academic competence winning her first-class Honours in Modern History and Political Economy. In the second year she won the Fulbright Scholarship to Sweet Briar College in Georgia. This was a seminal experience, bringing Jo face to face with segregation.

Coming from an impoverished, hungry Britain where rationing had been in place for 12 years, she was appalled at the profligacy and food waste in America. She worked as a waitress to earn pocket money and refused to remove diners' plates until all the food was eaten up. Thirty years later at the end of a reunion dinner at Sweet Briar, all her friends presented her with their clean plates—she had made her point and it had stuck!

Returning to St Andrews, she picked up where she had left off except that she made a new friend: me. Together every day, we discussed economics at a tea shop in Bell Street. The conversation naturally came around to what we would write for our degree theses. Mine was to be on sugar beet and the future for the beet factory ten miles down the road. Jo's was on the developing impact of the motorcar on our cities, health and economy, and our increasing dependence on imported fuels. It was an inspired piece of work, way ahead of its time, and led to the offer of a lectureship. She accepted the post, handing over presidency of the student union which she had undertaken in her fourth year.

By this time, she had consented to marry me and travel with me wherever work took us, just as her mother had done with her father. Seria in Brunei was our first marital home and where we had our first son, Douglas. Before he was born Jo taught in the local schools, enjoying the experience and the talent of her mostly Chinese pupils, but worried about the absence of Brunei children. Whether this was caused by financial problems or the overlay of the Catholic Church, she felt it was an opportunity missed. Her fascination with China began while we were in Brunei, stimulated by a three-week vacation in Hong Kong.

After getting to grips with the mystery of the silent and mysterious East, Jo was pitchforked into the mayhem of Nigeria. Roads crammed with cars, horns blazing, radios blasting and street merchants running alongside, hassling you to buy. It was moving upstream on a fast tide, jostling happily with boisterous and noisy Nigerians. Perhaps because Calcutta was in her roots, she loved it, and she celebrated by having our second son, Paul, who, like his mother, loved the place. It was his second home.

In a moment of madness, having seen the rich produce gardens in the middle of the country, Jo proposed a vegetable business to an optimistic tin miner, Malcolm Murrey. Together they negotiated a deal with a service-station owner whose site stood at the entry to the wealthy suburb of Lagos. The station, till then a run-of-the-mill operation, turned overnight into a centre of frenetic activity as expatriate wives queued for fresh vegetables, regardless of price. The dealer insisted you bought petrol before you could buy vegetables, so sales went through the roof. He was soon well on his way to being Dealer of the Year!

Malcolm the miner and his wife, Yvonne, were ex-military and had served with the British Army in Nigeria. A close friend of General Gowon, Malcolm was knowledgeable and wise about military government activities, and his advice was worth listening to. Jo's enterprise had cemented valuable friendships, leaving two Nigerians in Lagos wealthier in the process. For her the dividends were not counted in dollars—the financial advantage was for the players anyway—but in the widening range of her friends and the satisfaction of having delivered.

After we left Nigeria and went east again to Thailand, Jo caught up with some of her Cheltenham friends. A few who had royal titles invited her to meet the Queen—a picture of the party that day hangs on the sitting-room wall. Jo spent our four years in Thailand immersed in the country's history and that of its neighbours, Cambodia and Myanmar. On a visit to the latter we discovered the megastore that her step-uncles had built in the centre of Rangoon.

Our third son spent his early childhood in Thailand. Michael excelled in swimming and to our delight swam the backstroke leg for the Thai international team at the World Junior Championships in Singapore, and won a gold medal for the six-and-under relay. The magnet of the East had now worked its special magic on three generations.

Our final move east was to Australia, mecca of sport, where our two elder sons later settled and built their families. Australia felt like a breathing space after 20-odd years in the tropics, but it started with a tragedy when the chairman's wife, an Australian Olympic swimmer, died the night after her welcome home party. Jo realised the impact on the close family which was Shell, and played her part in restoring morale after the traumatic event. The widowed husband still speaks fondly of her help.

When a friend invited Jo to help Liberal Party supporters stuff envelopes, she went along, interested to see politics at work in Melbourne society. Collecting her piles of papers, she sat down to begin. After a few minutes the woman in charge tapped her on the shoulder and said, 'Thank you for coming—we need you. We have no national policy on abortion. Could you write one for us?' Jo never ducked a challenge and this was no exception, even though it was only a few days until the state conference at which the policy was to be tabled.

Surrounded by books—including the Bible—she worked three long days and deep into each night to have the paper ready for the conference, where it passed without amendment and was forwarded to the national conference for adoption. A month later it was unanimously accepted. This was legendary stuff.

When it was time to come back to Britain, the Royal Society of Arts was the first to secure Jo's services and she set about disentangling the knotted affairs of Benjamin Franklin House. In her Sweet Briar days she had, faced with the issues of segregation, sought solace from Franklin's life and writings, so it was a challenge she was determined to meet.

Her reputation as a social force for good soon saw her active in the East End of London. Schools there were unable to take advantage of high-quality teacher training because there was a lack of teacher accommodation. Jo had been there before; she knew a source of money not to be identified, but available. With the skills she had acquired on the babies' home project, she found a site for a residential block and had it open in time for the new school year. The infusion of enthusiastic young teachers had an immediate positive effect on local schools' performances.

Canterbury Christchurch College, soon to be a university, was among the partners in the initiative to improve education in the East End. Seeing the improvement Jo had orchestrated in teacher training, they invited her to join their board, and in time to become deputy chair. She was attracted because the university was attuned to the problems of everyday life and contributed to support services, including policing and nursing. Jo believed in life-long development and learning, and she achieved it.

*   *   *   *   *

A new interest emerged when Jo was invited to join The Past Overseers Society of St Margarets & St John, Westminster. This really intrigued her. The Overseers of the Poor, people of substance drawn from the local churches, were empowered by Queen Elizabeth I to remove vagrants from the City and return them to their place of origin. Jo and her friend Anne Lewis, with help from another friend Fiona Rogers, set out to write a history of the overseers. Their successors were no longer required to deal with vagrants, but this did not deter regular celebrations of their past achievements and the society continues, unburdened by such onerous responsibility. The book painted a picture of life and public concerns in Elizabethan times, drawing comparisons between then and now. The role of the Church, fear of plague, protection of citizens and social responsibility are not uncommon themes today.

Right up to her short illness and her passing, Jo was a stimulating person to be with: intellectually active, drawing on the widest range of international experience, and concerned about society. Our partnership was open and equal. During our long life together there were multiple discussions but never real disagreements; accommodation was in both our natures.

Perhaps the best picture of Jo is to be found in the words of our son, Douglas. From her earliest days Jo was in perpetual motion. As a result, he said, everyone held in their minds an image of her in full voice, with a joke, or a fiercely held opinion, or an insight, or a point of view—quite often a contrary point of view, but always a memorable and colourful one. We all had our pieces of that jigsaw, which made up the larger-than-life picture that was Jo Reid. Just like those madcap jigsaws she insisted we did every Christmas.

As she died holding my hand, she said pointedly, 'I have left them all alive!' My duty and my responsibility were clear, as they always had been.

Community Involvement

I firmly believe that leaders of large corporations that interact with citizens across the country and beyond should take active positions in public life. It is especially important that those running technical operations concern themselves with the establishments responsible for training the engineers, technologists and scientists of the future. Current leaders need to address the preparation and development of their successors.

My own experience with universities, colleges and schools has broadened my horizons, as has my involvement with a wide variety of public projects and activities. The opportunity to extend my range of interests by participating in the social, intellectual and charitable activities of my host and native countries has not been a burden but a blessing, and is made doubly enjoyable and rewarding when my background, experience or contacts enable me to make a lasting contribution.

Foundation for Management Education
Chairman 1986–2003

The Foundation evolved from meetings of senior businessmen who were concerned that Britain was failing to prepare graduates adequately for business. Spurred on by the flocks of MBAs who were making their mark in the US, the group put together a structure supported by experienced businessman Jim Roxborough, and asked me to chair the meetings.

Our first challenge was to define what should be in a business education programme's curriculum. There was a consensus against book learning and structured exams, and strong support for what we called management competences—essential business skills—to be defined. Having identified them, we would work out how they could best be acquired. We hired brilliant young executive Andrew Summers to advance this new approach. The concept of management competences caught on and soon educators were building it into their courses. A mini-revolution began and a coterie of supporters drawn from all parts of industry maintained its momentum.

St Edward's School
Chairman of Governors 1987–2016

The Industrial Society
Council Chairman 1993–1997

Founded in 1918 as the Boys' Welfare Association, the society switched to its current name in 1965. It played an active role in the 1970s when trade union, management and government formed joint economic development committees and wages councils to regulate working hours and conditions. But such divided authority and shared debating had no place in Margaret Thatcher's 'direct action' philosophy. I was asked to become chairman during her premiership when the society was caught in a maelstrom of change. Olympic sailor Tony Morgan was then chief executive and was implementing a programme of rejuvenation, improving industrial and commercial management relationships. Unfortunately, as often happens, the programme did not continue after he left, though I consoled myself that his achievements would be carried forward by the people he had worked with.

Contributors' efforts are inevitably limited by their terms of office, which is a real issue in Britain because much of the time spent on these important endeavours is voluntary, and given at the expense of other interests.

Robert Gordon University
First Chancellor 1993–2004

Robert Gordon of Aberdeen made his fortune in the mid-eighteenth century from trading with Baltic ports. His operations were centred in Danzig. Concerned at the poverty and living conditions of the young in his home town, he established Robert Gordon Hospital, which provided a basic education and with it, the opportunity to find employment.

During the nineteenth century, Robert Gordon's and other institutes merged to become Robert Gordon's Technical College. It became Robert Gordon's Institute of Technology in 1965, its research activity grew and it became a degree-granting institution and finally a university in 1992. Shell had recruited from the university and used its staff and facilities to train offshore personnel in survival skills. I have seen a rescue from a ditched helicopter in which that training undoubtedly saved lives.

In 1992 I accepted an invitation to be chancellor, happy to have an opportunity to give something back to a city that had supported our activities in the North Sea and welcomed colleagues and their families. In addition, my father, having been a Gordon Highlander, had strong personal links with Aberdeen.

In accepting the role I made it clear that I wanted a modern organization without courts and rectors but with a straightforward executive team. I had in mind the man to head the executive: Sir Ian Wood, an Aberdeen business leader, was an ideal fit and his appointment assured progress.

The university's successful pharmacy department attracted overseas students, some of whose parents were already involved in pharmacy. The nursing faculty was also hugely useful to healthcare in the north of Scotland, where doctors were thinly spread. Nurses needed to be able to handle a broad range of ailments. I tried without success to persuade my colleagues to design a qualification that would enable nurses to take on greater responsibility, but they set the intellectual criteria too high. They felt that the provision of education and training for such a practical purpose was outside their remit, and saw the proposition as a potential dilution of academic standards rather than as meeting real human needs.

By the end of the '90s, the new university was settling down and its graduation ceremonies gave the public insight into the place. Student activities especially in sports were well supported, and clubs and associations flourished, as did the students' union. Robert Gordon would have been pleased with the development of his brainchild and the opportunities his vision has opened up for many young people, irrespective of their financial circumstances. He would also have been delighted with the promotion of Sir Ian Wood, a true Aberdonian, to the chancellorship in 2005.

Foundation for Young Musicians
Chairman 1994–

I inherited board membership of the foundation from John Wilson when I became chairman of London Electricity. The organization encourages young Londoners with an interest in and talent for music to take advantage of the facilities at Morley College Centre for Young Musicians (CYM). CYM's young people give performances across the city as members of the London Schools Symphony Orchestra, enabling the wider community to enjoy their talent. The college gives more than 2,600 young people a year high-quality instrumental and vocal training, and aims to ensure that no young person is turned away because of lack of funds. Students aged eight to 19 from all over London are identified—often by their schools or borough music services—for their potential, rather than for their existing level of achievement, and given the opportunity to progress.

The chairman was a former Cabinet secretary who famously coined the phrase 'economical with the truth'. He was. He asked me to stand in for a couple of months and still remains absent over two decades later. But the position, alongside truly professional colleagues, is a pleasure to hold. An arrangement negotiated with the Guildhall School of Music has helped secure the centre's future.

Benjamin Franklin House
Chairman Emeritus 1997–2007

Jo, on the committee of Benjamin Franklin House, was concerned at its lack of direction and persuaded—or rather ordered—me to become chairman of the Friends.

Franklin arrived in London in August 1757 with his son, William, and two slaves, Peter and King. Within a couple of days Mrs Stevenson of Craven Street accepted the four as boarders, one room each for Franklin and William and the attic for the two 'bonded servants'—then a more acceptable term in London! Franklin stayed until 1762 and returned in 1775. His interest in science and membership of the Royal Society forged friendships and opened doors into intellectual circles. His shortened stay and subsequent appointment as ambassador in France, where he was charged with gaining French support for American Independence, rather damaged his reputation in Britain, but his strong personal friendships survived.

Restoring the house was a complex physical proposition, starting with the walls, which needed the support of an all-embracing steel corset. While work was being carried out inside, I received an urgent call from the builders, who had found human bones under the floor and—weirdly—lobster shells, complete with legs and claws. The imagination ran wild but the historians soon put us straight. The landlady's son-in-law had been a surgeon. He had conducted demonstrations on cadavers in the basement, and his thesis for membership of the Royal Society was on the lobster's lymphatic system.

It was not the structure of the board that made the work at Benjamin Franklin House challenging; it was the nature of the members. Mary Bessborough was the granddaughter of A. J. Drexel from the esteemed banking family. She had married an English earl and was now widowed. Wealthy, governess-educated and beautiful, she was generous but could be taken advantage of. Bills were run up without any material progress being made, and attempts to control expenditure were often fruitless.

We needed a chief executive who was more attuned to the challenge of budgeting. At Jo's suggestion we appointed American Dr Márcia Balisciano, who took control and managed both Mary and the restoration brilliantly, with excellent help from Graham Nixon. Secretary Hamid Rashidmanesh, a calm, efficient Iranian designated by his law firm as a charitable donation to the project, successfully forestalled the legal challenges that occasionally arose.

This was one of the best experiences of my working life. The director managed a sometimes-disruptive, albeit creative, group with strength, determination and charm, patiently and courteously heading off maverick diversions; she raised funds tirelessly and looked after one of America's designated treasures. It was a sterling performance. The well-ordered life of my commercial businesses had provided much excitement, but not the kind of frustration we encountered here. In the world of charitable institutions, achievements are hard won, frustrations are many and those who become involved richly deserve support.

Learning Through Landscapes
Chair of Trustees 2000–2019

This energetic initiative was inspired by David Attenborough. It helps primary schools to use their outdoor premises creatively, introducing the children in a practical way to the natural world and the rewards of cultivation. Jo invited me to join the governing committee and, just as she did, I found the work involving and rewarding.

As an introduction to science and horticulture, Project Fruitful subsidised the planting of fruit trees in school grounds, bringing blossom in spring and harvest in autumn. This was followed by a study of pollination, looking at the work of insects, bees, butterflies and moths, and explaining how different flowers contribute different flavours to honey. The scheme reached as far as northern Scotland.

Making connections between the created and natural worlds presents fascinating challenges and rewards for both teacher and student, and I am convinced helps to develop better-informed, knowledgeable citizens.

Conservatoire for Dance and Drama
Chairman 2001–2011

Richard Attenborough knew how to tie you in, and BAFTA was just the start. His next assignment was to ensure that interest in dance and drama was spread across the country and its varied communities. I was corralled to preside over a conglomerate of schools and academies, our purpose being to ensure that the government was aware of their work and their financial needs, and in return we provided assurance that their processes and accounting were up to standard.

My individual association was with the London Contemporary Dance School, situated just off Marylebone Road and under the direction of Veronica Lewis. Veronica, a former dancer, was a first-class director and managed a school of outstanding dancers.

Edinburgh Business School
Chair 2001–2015

Edinburgh Business School is the graduate business school of Heriot-Watt University. Professor Keith Lumsden, the school's head, and Alick Kitchen, director of studies, took on the challenge of building the school into a business. The Caribbean and Mauritius were good sources of students; South America, India, the Middle and Far East also offered possibilities which they pursued along with a lucrative London market. Keith was determined to maintain the highest academic standards, and as a result the school's quality was recognized in the industry, as were its financial results. Accrued reserves reached well over £20 million, which was unusual for a relatively small institution.

In around 2010, tension between the university and the school arose centred on real-estate development. This was no ordinary piece of real estate, however; this was Panmure House, the eighteenth-century home of economist and philosopher Adam Smith. Keith, with his strong American contacts, was confident that he had the finance in place and that the property would be acquired. The court became involved and what should have been a straightforward transaction became a protracted affair, but eventually the school managed to take the house on. It was an excellent project for us, its political connections in business education and commercial finance well-placed to boost our growth and reputation at home and abroad.

An attractive conference centre is now in place on the Royal Mile. By the time the project was completed, Keith was close to, if not over, retirement age, but I was pleased to be able to arrange for him to be the first warden of Panmure House just before I left. The opening ceremony in late 2018 was led by the Rt Hon Gordon Brown, former prime minister and United Nations special envoy for global education, who—like Adam Smith—had his roots in Kirkcaldy.

At my honorary graduation in 2018 the new principal of the university was excited about the opportunities it presented. The best laid schemes o' mice an' men gang aft a-gley—happily this time, although they tried, they did not!

The Thatcher Initiative in Eastern Europe

Mrs Thatcher was concerned that universities in countries emerging from communist influence would not have the freedom of thought that enriches student development and works in turn to improve the country's economy. She decided to tackle this and asked me to help. The plan was to interface the relevant nation's universities with our own, hoping by that means to instigate a kind of revolution.

First we had to find universities that wanted to be involved, and then look for positive partnerships within them. We concentrated on economics and business studies, with the primary emphasis on business. Warsaw University was the first candidate, and we selected Nottingham University as its opposite number. Among the staff of our Shell offices in Poland was a colleague from my Nigerian days. He was used to my pursuit of the outlandish, though he did admit that from time to time the schemes worked out well. This one would be hard, but through his excellent contacts we made progress in understanding what was currently being taught at the university. It was depressing; the syllabus for the 'business' faculty was based on the government's regulatory legislation manual, and this is what the students were examined in. Entrepreneurism was absent from the curriculum, as were the other essential commercial and financial tools that build a prosperous economy.

We introduced Nottingham and a lively programme of free-market economics into this well-ordered hive of unproductivity. Our proposed curriculum triggered a mass exodus of the resident staff, but the students loved the new model and applied themselves, helping to bring about a successful peaceful revolution. The academics who had not fled gradually persuaded many of their former colleagues to return—but to a different world.

Disengaging from the past was sometimes a dangerously stormy process, but the Polish leaders knew we were on the right track and supported the renaissance. Reading the university's website now and seeing the network of relationships it has subsequently built, it is clear that the revolution goes on, and that this is now a twenty-first-century institution.

My next assignment was Bratislava in Slovakia. After Warsaw, I expected to find another faculty in the frozen grip of communist ideology and braced myself for a long, uphill battle.

Bratislava is beautiful. It bestrides the Danube, one side nestling in the foothills of the Little Carpathian mountains. With a spectacular Opera House, opened as the city theatre in 1886, the place had a strange flexibility and fluidity about it. This was underlined at a dinner that the Slovakian refinery's managing director kindly organized for me. He was clearly distressed by the world he found himself in. His operation, staff and training schemes were exceptional, but they were hobbled by a sadly impoverished commercial environment. Margins were minimal and strictly controlled. Staff wages were not quite at starvation level, but they were certainly far lower than the European going rate. With the new freedom of movement, his staff were leaving, moving elsewhere where there were significant financial advantages. What had started as a trickle was becoming a torrent, leaving him effectively as an overseas training centre for refinery executives and technicians. He was not a happy man and I was almost certain he would be the next to go.

With these thoughts in mind I went to investigate Bratislava University. Commerce and business, they stressed, were the new opportunity. Walls had come down, strait jackets, if they ever existed, had been cast aside. They saw their opera house as a huge opportunity; tourism could be built around that. The mountains offered further potential with summer walking and winter skiing—and then there was the sea. They were not building a university, they were building a multifaceted enterprise with strong international attraction.

The challenge at this seat of learning was certainly a different one—to get them back closer to education and theory, but their appetite for real commerce and real business would have made Mrs Thatcher a happy woman.